This is a printed compilation for people that enjoy using and working with printed manuals. The information in this compilation is available for free in PDF format directly from Raspberry Pi. This manual is printed in accordance with their CC BY-ND license. This is a third party printing of their documentation by DienstNet LLC. As an extension, all parts of this compilation not covered by the Raspberry Pi license are also provided under the same CC-BY-ND copyright by DienstNet LLC 2023.

Compilation Contents
Raspberry Pi Pico C/C++ SDK
https://datasheets.raspberrypi.com/pico/raspberry-pi-pico-c-sdk.pdf

ISBN 978-1-365-38152-2

This page was intentionally left blank.

Raspberry Pi Pico C/C++ SDK

Libraries and tools for
C/C++ development on
RP2040 microcontrollers

Colophon

Copyright © 2020-2022 Raspberry Pi Ltd (formerly Raspberry Pi (Trading) Ltd.)

The documentation of the RP2040 microcontroller is licensed under a Creative Commons [Attribution-NoDerivatives 4.0 International](#) (CC BY-ND).

build-date: 2022-11-30
build-version: 3a2defe-clean

About the SDK

Throughout the text "the SDK" refers to our [Raspberry Pi Pico SDK](#). More details about the SDK can be found throughout this book. Source code included in the documentation is Copyright © 2020-2022 Raspberry Pi Ltd (formerly Raspberry Pi (Trading) Ltd.) and licensed under the [3-Clause BSD](#) license.

Legal disclaimer notice

TECHNICAL AND RELIABILITY DATA FOR RASPBERRY PI PRODUCTS (INCLUDING DATASHEETS) AS MODIFIED FROM TIME TO TIME ("RESOURCES") ARE PROVIDED BY RASPBERRY PI LTD ("RPL") "AS IS" AND ANY EXPRESS OR IMPLIED WARRANTIES, INCLUDING, BUT NOT LIMITED TO, THE IMPLIED WARRANTIES OF MERCHANTABILITY AND FITNESS FOR A PARTICULAR PURPOSE ARE DISCLAIMED. TO THE MAXIMUM EXTENT PERMITTED BY APPLICABLE LAW IN NO EVENT SHALL RPL BE LIABLE FOR ANY DIRECT, INDIRECT, INCIDENTAL, SPECIAL, EXEMPLARY, OR CONSEQUENTIAL DAMAGES (INCLUDING, BUT NOT LIMITED TO, PROCUREMENT OF SUBSTITUTE GOODS OR SERVICES; LOSS OF USE, DATA, OR PROFITS; OR BUSINESS INTERRUPTION) HOWEVER CAUSED AND ON ANY THEORY OF LIABILITY, WHETHER IN CONTRACT, STRICT LIABILITY, OR TORT (INCLUDING NEGLIGENCE OR OTHERWISE) ARISING IN ANY WAY OUT OF THE USE OF THE RESOURCES, EVEN IF ADVISED OF THE POSSIBILITY OF SUCH DAMAGE.

RPL reserves the right to make any enhancements, improvements, corrections or any other modifications to the RESOURCES or any products described in them at any time and without further notice.

The RESOURCES are intended for skilled users with suitable levels of design knowledge. Users are solely responsible for their selection and use of the RESOURCES and any application of the products described in them. User agrees to indemnify and hold RPL harmless against all liabilities, costs, damages or other losses arising out of their use of the RESOURCES.

RPL grants users permission to use the RESOURCES solely in conjunction with the Raspberry Pi products. All other use of the RESOURCES is prohibited. No licence is granted to any other RPL or other third party intellectual property right.

HIGH RISK ACTIVITIES. Raspberry Pi products are not designed, manufactured or intended for use in hazardous environments requiring fail safe performance, such as in the operation of nuclear facilities, aircraft navigation or communication systems, air traffic control, weapons systems or safety-critical applications (including life support systems and other medical devices), in which the failure of the products could lead directly to death, personal injury or severe physical or environmental damage ("High Risk Activities"). RPL specifically disclaims any express or implied warranty of fitness for High Risk Activities and accepts no liability for use or inclusions of Raspberry Pi products in High Risk Activities.

Raspberry Pi products are provided subject to RPL's [Standard Terms](#). RPL's provision of the RESOURCES does not expand or otherwise modify RPL's [Standard Terms](#) including but not limited to the disclaimers and warranties expressed in them.

Table of contents

Colophon 1
    Legal disclaimer notice 1
1. About the SDK 6
    1.1. Introduction 6
    1.2. Anatomy of a SDK Application 6
2. SDK architecture 9
    2.1. The Build System 9
    2.2. Every Library is an INTERFACE 10
    2.3. SDK Library Structure 11
        2.3.1. Higher-level Libraries 11
        2.3.2. Runtime Support (pico_runtime, pico_standard_link) 11
        2.3.3. Hardware Support Libraries 12
        2.3.4. Hardware Structs Library 13
        2.3.5. Hardware Registers Library 14
        2.3.6. TinyUSB Port 15
    2.4. Directory Structure 15
        2.4.1. Locations of Files 16
    2.5. Conventions for Library Functions 17
        2.5.1. Function Naming Conventions 17
        2.5.2. Return Codes and Error Handling 18
        2.5.3. Use of Inline Functions 19
        2.5.4. Builder Pattern for Hardware Configuration APIs 19
    2.6. Customisation and Configuration Using Preprocessor variables 20
        2.6.1. Preprocessor Variables via Board Configuration File 21
        2.6.2. Preprocessor Variables Per Binary or Library via CMake 21
    2.7. SDK Runtime 22
        2.7.1. Standard Input/Output (stdio) Support 22
        2.7.2. Floating-point Support 22
        2.7.3. Hardware Divider 25
    2.8. Multi-core support 26
    2.9. Using C++ 26
    2.10. Next Steps 27
3. Using programmable I/O (PIO) 28
    3.1. What is Programmable I/O (PIO)? 28
        3.1.1. Background 28
        3.1.2. I/O Using dedicated hardware on your PC 28
        3.1.3. I/O Using dedicated hardware on your Raspberry Pi or microcontroller 28
        3.1.4. I/O Using software control of GPIOs ("*bit-banging*") 29
        3.1.5. Programmable I/O Hardware using FPGAs and CPLDs 30
        3.1.6. Programmable I/O Hardware using PIO 30
    3.2. Getting started with PIO 31
        3.2.1. A First PIO Application 31
        3.2.2. A Real Example: WS2812 LEDs 35
        3.2.3. PIO and DMA (A Logic Analyser) 43
        3.2.4. Further examples 48
    3.3. Using PIOASM, the PIO Assembler 48
        3.3.1. Usage 48
        3.3.2. Directives 49
        3.3.3. Values 50
        3.3.4. Expressions 50
        3.3.5. Comments 50
        3.3.6. Labels 51
        3.3.7. Instructions 51
        3.3.8. Pseudoinstructions 51
        3.3.9. Output pass through 52

- 3.3.10. Language generators ... 52
- 3.4. PIO Instruction Set Reference ... 57
  - 3.4.1. Summary ... 58
  - 3.4.2. JMP ... 58
  - 3.4.3. WAIT ... 59
  - 3.4.4. IN ... 60
  - 3.4.5. OUT ... 61
  - 3.4.6. PUSH ... 62
  - 3.4.7. PULL ... 63
  - 3.4.8. MOV ... 64
  - 3.4.9. IRQ ... 66
  - 3.4.10. SET ... 67
- 4. Library documentation ... 68
- Appendix A: App Notes ... 69
  - Attaching a 7 segment LED via GPIO ... 69
    - Wiring information ... 69
    - List of Files ... 69
    - Bill of Materials ... 71
  - DHT-11, DHT-22, and AM2302 Sensors ... 72
    - Wiring information ... 72
    - List of Files ... 73
    - Bill of Materials ... 75
  - Attaching a 16x2 LCD via TTL ... 75
    - Wiring information ... 76
    - List of Files ... 76
    - Bill of Materials ... 79
  - Attaching a microphone using the ADC ... 80
    - Wiring information ... 81
    - List of Files ... 81
    - Bill of Materials ... 82
  - Attaching a BME280 temperature/humidity/pressure sensor via SPI ... 83
    - Wiring information ... 83
    - List of Files ... 83
    - Bill of Materials ... 88
  - Attaching a MPU9250 accelerometer/gyroscope via SPI ... 88
    - Wiring information ... 88
    - List of Files ... 89
    - Bill of Materials ... 92
  - Attaching a MPU6050 accelerometer/gyroscope via I2C ... 92
    - Wiring information ... 93
    - List of Files ... 93
    - Bill of Materials ... 96
  - Attaching a 16x2 LCD via I2C ... 96
    - Wiring information ... 96
    - List of Files ... 96
    - Bill of Materials ... 100
  - Attaching a BMP280 temp/pressure sensor via I2C ... 100
    - Wiring information ... 100
    - List of Files ... 101
    - Bill of Materials ... 106
  - Attaching a LIS3DH Nano Accelerometer via i2c ... 106
    - Wiring information ... 106
    - List of Files ... 107
    - Bill of Materials ... 110
  - Attaching a MCP9808 digital temperature sensor via I2C ... 110
    - Wiring information ... 110
    - List of Files ... 110
    - Bill of Materials ... 113
  - Attaching a MMA8451 3-axis digital accelerometer via I2C ... 114
    - Wiring information ... 114

- List of Files 114
- Bill of Materials 117
- Attaching an MPL3115A2 altimeter via I2C 117
  - Wiring information 118
  - List of Files 118
  - Bill of Materials 122
- Attaching an OLED display via I2C 122
  - Wiring information 124
  - List of Files 124
  - Bill of Materials 130
- Attaching a PA1010D Mini GPS module via I2C 130
  - Wiring information 130
  - List of Files 131
  - Bill of Materials 134
- Attaching a PCF8523 Real Time Clock via I2C 134
  - Wiring information 134
  - List of Files 134
  - Bill of Materials 138
- Appendix B: SDK configuration 139
  - Configuration Parameters 140
- Appendix C: CMake build configuration 147
  - Configuration Parameters 147
    - Control of binary type produced (advanced) 148
- Appendix D: Board configuration 149
  - Board Configuration 149
    - The Configuration files 149
    - Building applications with a custom board configuration 151
    - Available configuration parameters 151
- Appendix E: Building the SDK API documentation 152
- Appendix F: SDK release history 153
  - Release 1.0.0 (20/Jan/2021) 153
  - Release 1.0.1 (01/Feb/2021) 153
    - Boot Stage 2 153
  - Release 1.1.0 (05/Mar/2021) 153
    - Backwards incompatibility 154
  - Release 1.1.1 (01/Apr/2021) 154
  - Release 1.1.2 (07/Apr/2021) 154
  - Release 1.2.0 (03/Jun/2021) 154
    - New/improved Board headers 154
    - Updated TinyUSB to 0.10.1 154
    - Added CMSIS core headers 154
    - API improvements 155
    - General code improvements 156
    - SVD 157
    - pioasm 157
    - RTOS interoperability 157
    - CMake build changes 157
    - Boot Stage 2 157
  - Release 1.3.0 (02/Nov/2021) 157
    - Updated TinyUSB to 0.12.0 157
    - New Board Support 157
    - Updated SVD, `hardware_regs`, `hardware_structs` 158
    - Behavioural Changes 159
    - Other Notable Improvements 159
    - CMake build 161
    - pioasm 161
    - elf2uf2 161
  - Release 1.3.1 (18/May/2022) 161
    - New Board Support 161
    - Notable Library Changes/Improvements 162

	Build	162
	pioasm	163
	elf2uf2	163
Release 1.4.0 (30/Jun/2022)		163
	New Board Support	163
	Wireless Support	163
	Notable Library Changes/Improvements	164
	Build	166
Appendix G: Documentation release history		167

Chapter 1. About the SDK

1.1. Introduction

The SDK (Software Development Kit) provides the headers, libraries and build system necessary to write programs for RP2040-based devices such as Raspberry Pi Pico in C, C++ or Arm assembly language.

The SDK is designed to provide an API and programming environment that is familiar both to non-embedded C developers and embedded C developers alike. A single program runs on the device at a time with a conventional `main()` method. Standard C/C++ libraries are supported along with APIs for accessing RP2040's hardware, including DMA, IRQs, and the wide variety fixed function peripherals and PIO (Programmable IO).

Additionally the SDK provides higher level libraries for dealing with timers, USB, synchronization and multi-core programming, along with additional high level functionality built using PIO such as audio. These libraries should be comprehensive enough that your application code rarely, if at all, needs to access hardware registers directly. However, if you do need or prefer to access the raw hardware, you will also find complete and fully-commented register definition headers in the SDK. There's no need to look up addresses in the datasheet.

The SDK can be used to build anything from simple applications, full fledged runtime environments such as MicroPython, to low level software such as RP2040's on-chip bootrom itself.

Looking to get started?

This book documents the SDK APIs, explains the internals and overall design of the SDK, and explores some deeper topics like using the PIO assembler to build new interfaces to external hardware. For a quick start with setting up the SDK and writing SDK programs, Getting started with Raspberry Pi Pico is the best place to start.

1.2. Anatomy of a SDK Application

Before going completely depth-first in our traversal of the SDK, it's worth getting a little breadth by looking at one of the SDK examples covered in Getting started with Raspberry Pi Pico, in more detail.

Pico Examples: https://github.com/raspberrypi/pico-examples/blob/master/blink/blink.c

```c
/**
 * Copyright (c) 2020 Raspberry Pi (Trading) Ltd.
 *
 * SPDX-License-Identifier: BSD-3-Clause
 */

#include "pico/stdlib.h"

int main() {
#ifndef PICO_DEFAULT_LED_PIN
#warning blink example requires a board with a regular LED
#else
    const uint LED_PIN = PICO_DEFAULT_LED_PIN;
    gpio_init(LED_PIN);
    gpio_set_dir(LED_PIN, GPIO_OUT);
    while (true) {
        gpio_put(LED_PIN, 1);
        sleep_ms(250);
```

```
19          gpio_put(LED_PIN, 0);
20          sleep_ms(250);
21      }
22 #endif
23 }
```

This program consists only of a single C file, with a single function. As with almost any C programming environment, the function called `main()` is special, and is the point where the language runtime first hands over control to your program, after doing things like initialising static variables with their values. In the SDK the `main()` function does not take any arguments. It's quite common for the `main()` function not to return, as is shown here.

> ℹ️ **NOTE**
>
> The return code of `main()` is ignored by the SDK runtime, and the default behaviour is to hang the processor on exit.

At the top of the C file, we include a header called `pico/stdlib.h`. This is an umbrella header that pulls in some other commonly used headers. In particular, the ones needed here are `hardware/gpio.h`, which is used for accessing the general purpose IOs on RP2040 (the `gpio_xxx` functions here), and `pico/time.h` which contains, among other things, the `sleep_ms` function. Broadly speaking, a library whose name starts with `pico` provides high level APIs and concepts, or aggregates smaller interfaces; a name beginning with `hardware` indicates a thinner abstraction between your code and RP2040 on-chip hardware.

So, using mainly the `hardware_gpio` and `pico_time` libraries, this C program will blink an LED connected to GPIO25 on and off, twice per second, forever (or at least until unplugged). In the directory containing the C file (you can click the link above the source listing to go there), there is one other file which lives alongside it.

Directory listing of pico-examples/blink

```
blink
├── blink.c
└── CMakeLists.txt

0 directories, 2 files
```

The second file is a *CMake* file, which tells the SDK how to turn the C file into a binary application for an RP2040-based microcontroller board. Later sections will detail exactly what CMake is, and why it is used, but we can look at the contents of this file without getting mired in those details.

Pico Examples: https://github.com/raspberrypi/pico-examples/blob/master/blink/CMakeLists.txt

```
 1 add_executable(blink
 2         blink.c
 3         )
 4
 5 # pull in common dependencies
 6 target_link_libraries(blink pico_stdlib)
 7
 8 # create map/bin/hex file etc.
 9 pico_add_extra_outputs(blink)
10
11 # add url via pico_set_program_url
12 example_auto_set_url(blink)
```

The `add_executable` function in this file declares that a program called `blink` should be built from the C file shown earlier. This is also the target name used to build the program: in the `pico-examples` repository you can say `make blink` in your build directory, and that name comes from *this* line. You can have multiple executables in a single project, and the `pico-examples` repository is one such project.

The `target_link_libraries` is pulling in the SDK functionality that our program needs. If you don't ask for a library, it doesn't appear in your program binary. Just like `pico/stdlib.h` is an umbrella header that includes things like `pico/time.h` and `hardware/gpio.h`, `pico_stdlib` is an umbrella *library* that makes libraries like `pico_time` and `hardware_gpio` available to your build, so that those headers can be included in the first place, and the extra C source files are compiled and linked. If you need less common functionality, like accessing the DMA hardware, you can call those libraries out here (e.g. listing `hardware_dma` before or after `pico_stdlib`).

We could end the CMake file here, and that would be enough to build the `blink` program. By default, the build will produce an ELF file (executable linkable format), containing all of your code and the SDK libraries it uses. You can load an ELF into RP2040's RAM or external flash through the Serial Wire Debug port, with a debugger setup like `gdb` and `openocd`. It's often easier to program your Raspberry Pi Pico or other RP2040 board directly over USB with BOOTSEL mode, and this requires a different type of file, called UF2, which serves the same purpose here as an ELF file, but is constructed to survive the rigours of USB mass storage transfer more easily. The `pico_add_extra_outputs` function declares that you want a UF2 file to be created, as well as some useful extra build output like disassembly and map files.

> **NOTE**
>
> The ELF file is converted to a UF2 with an internal SDK tool called `elf2uf2`, which is bootstrapped automatically as part of the build process.

The `example_auto_set_url` function is to do with how you are able to read this source file in *this* document you are reading right now, and click links to take you to the listing on GitHub. You'll see this on the `pico-examples` applications, but it's not necessary on your own programs. You are seeing how the sausage is made.

Finally, a brief note on the `pico_stdlib` library. Besides common hardware and high-level libraries like `hardware_gpio` and `pico_time`, it also pulls in components like `pico_standard_link` — which contains linker scripts and `crt0` for SDK — and `pico_runtime`, which contains code running between `crt0` and `main()`, getting the system into a state ready to run code by putting things like clocks and resets in a safe initial state. These are incredibly low-level components that most users will not need to worry about. The reason they are mentioned is to point out that they are ultimately *explicit dependencies* of your program, and you can choose not to use them, whilst still building against the SDK and using things like the `hardware` libraries.

Chapter 2. SDK architecture

RP2040 is a powerful chip, and in particular was designed with a disproportionate amount of system RAM for its point in the microcontroller design space. However it *is* an embedded environment, so RAM, CPU cycles and program space are still at a premium. As a result the tradeoffs between performance and other factors (e.g. edge case error handling, runtime vs compile time configuration) are necessarily much more visible to the developer than they might be on other, higher level platforms.

The intention within the SDK has been for features to just work out of the box, with sensible defaults, but also to give the developer as much control and power as possible (if they want it) to fine tune every aspect of the application they are building and the libraries used.

The next few sections try to highlight some of the design decisions behind the SDK: the how and the why, as much as the what.

 NOTE

> Some parts of this overview are quite technical or deal with very low-level parts of the SDK and build system. You might prefer to skim this section at first and then read it thoroughly at a later time, after writing a few SDK applications.

2.1. The Build System

The SDK uses CMake to manage the build. CMake is widely supported by IDEs (Integrated Development Environments), which can use a `CMakeLists.txt` file to discover source files and generate code autocomplete suggestions. The same `CMakeLists.txt` file provides a terse specification of how your application (or your project with many distinct applications) should be built, which CMake uses to generate a robust build system used by `make`, `ninja` or other build tools. The build system produced is customised for the platform (e.g. Windows, or a Linux distribution) and by any configuration variables the developer chooses.

Section 2.6 shows how CMake can set configuration defines for a particular program, or based on which RP2040 *board* you are building for, to configure things like default pin mappings and features of SDK libraries. These defines are listed in Appendix B, and Board Configuration files are covered in more detail in Appendix D. Additionally Appendix C describes CMake variables you can use to control the functionality of the build itself.

Apart from being a widely used build system for C/C++ development, CMake is fundamental to the way the SDK is structured, and how applications are configured and built.

Pico Examples: https://github.com/raspberrypi/pico-examples/blob/master/blink/CMakeLists.txt

```
1  add_executable(blink
2          blink.c
3          )
4
5  # pull in common dependencies
6  target_link_libraries(blink pico_stdlib)
7
8  # create map/bin/hex file etc.
9  pico_add_extra_outputs(blink)
10
11 # add url via pico_set_program_url
12 example_auto_set_url(blink)
```

Looking here at the blink example, we are defining a new executable `blink` with a single source file `blink.c`, with a single dependency `pico_stdlib`. We also are using a SDK provided function `pico_add_extra_outputs` to ask additional files to be produced beyond the executable itself (`.uf2`, `.hex`, `.bin`, `.map`, `.dis`).

The SDK builds an executable which is `bare metal`, i.e. it includes the entirety of the code needed to run on the device (other than floating point and other optimized code contained in the bootrom within RP2040).

`pico_stdlib` is an `INTERFACE` library and provides all of the rest of the code and configuration needed to compile and link the `blink` application. You will notice if you do a build of blink (https://github.com/raspberrypi/pico-examples/blob/master/blink/blink.c) that in addition to the single `blink.c` file, the inclusion of `pico_stdlib` causes about 40 other source files to be compiled to flesh out the blink application such that it can be run on RP2040.

2.2. Every Library is an INTERFACE

All libraries within the SDK are `INTERFACE` libraries. (Note this does not include the C/C++ standard libraries provided by the compiler). Conceptually, a CMake `INTERFACE` library is a collection of:

- Source files
- Include paths
- Compiler definitions (visible to code as `#defines`)
- Compile and link options
- Dependencies (on other `INTERFACE` libraries)

The `INTERFACE` libraries form a tree of dependencies, with each contributing source files, include paths, compiler definitions and compile/link options to the build. These are collected based on the libraries you have listed in your `CMakeLists.txt` file, and the libraries depended on by *those* libraries, and so on recursively. To build the application, each source file is compiled with the combined include paths, compiler definitions and options and linked into an executable according to the provided link options.

When building an executable with the SDK, all of the code for one executable, including the SDK libraries, is (re)compiled for *that* executable from source. Building in this way allows your build configuration to specify customised settings for those libraries (e.g. enabling/disabling assertions, setting the sizes of static buffers), on a *per-application* basis, at compile time. This allows for faster and smaller binaries, in addition of course to the ability to remove support for unwanted features from your executable entirely.

In the example `CMakeLists.txt` we declare a dependency on the (`INTERFACE`) library `pico_stdlib`. This `INTERFACE` library itself depends on other `INTERFACE` libraries (`pico_runtime`, `hardware_gpio`, `hardware_uart` and others). `pico_stdlib` provides all the basic functionality needed to get a simple application running and toggling GPIOs and printing to a UART, and the linker will garbage collect any functions you don't call, so this doesn't bloat your binary. We can take a quick peek into the directory structure of the `hardware_gpio` library, which our `blink` example uses to turn the LED on and off:

```
hardware_gpio
├── CMakeLists.txt
├── gpio.c
└── include
    └── hardware
        └── gpio.h
```

Depending on the `hardware_gpio` `INTERFACE` library in your application causes `gpio.c` to be compiled and linked into your executable, and adds the `include` directory shown here to your search path, so that a `#include "hardware/gpio.h"` will pull in the correct header in your code.

`INTERFACE` libraries also make it easy to aggregate functionality into readily consumable chunks (such as `pico_stdlib`), which don't directly contribute any code, but depend on a handful of lower-level libraries that do. Like a metapackage, this lets you pull in a group of libraries related to a particular goal without listing them all by name.

> **❗ IMPORTANT**
>
> SDK functionality is grouped into separate `INTERFACE` libraries, and each `INTERFACE` library contributes the code *and* include paths for that library. Therefore you must declare a dependency on the `INTERFACE` library you need directly (or indirectly through another `INTERFACE` library) for the header files to be found during compilation of your source file (or for code completion in your IDE).

> **ℹ NOTE**
>
> As all libraries within the SDK are `INTERFACE` libraries, we will simply refer to them as libraries or SDK libraries from now on.

2.3. SDK Library Structure

The full API listings are given in Chapter 4; this chapter gives an overview of how SDK libraries are organised, and the relationships between them.

There are a number of layers of libraries within the SDK. This section starts with the highest-level libraries, which can be used in C or C++ applications, and navigates all the way down to the `hardware_regs` library, which is a comprehensive set of hardware definitions suitable for use in Arm assembly as well as C and C++, before concluding with a brief note on how the TinyUSB stack can be used from within the SDK.

2.3.1. Higher-level Libraries

These libraries (`pico_xxx`) provide higher-level APIs, concepts and abstractions. The APIs are listed in High Level APIs. These may be libraries that have cross-cutting concerns between multiple pieces of hardware (for example the `sleep_` functions in `pico_time` need to concern themselves both with RP2040's timer hardware and with how processors enter and exit low power states), or they may be pure software infrastructure required for your program to run smoothly. This includes libraries for things like:

- Alarms, timers and time functions
- Multi-core support and synchronization primitives
- Utility functions and data structures

These libraries are generally built upon one or more underlying `hardware_` libraries, and often depend on each other.

> **ℹ NOTE**
>
> More libraries will be forthcoming in the future (e.g. - Audio support (via PIO), DPI/VGA/MIPI Video support (via PIO) file system support, SDIO support via (PIO)), most of which are available but not yet fully supported/stable/documented in the Pico Extras GitHub repository.

2.3.2. Runtime Support (pico_runtime, pico_standard_link)

These are libraries that bundle functionality which is common to most RP2040-based applications. APIs are listed in Runtime Infrastructure.

`pico_runtime` aggregates the libraries (listed in `pico_runtime`) that provide a familiar C environment for executing code, including:

- Runtime startup and initialization

- Choice of language level single/double precision floating point support (and access to the fast on-RP2040 implementations)
- Compact `printf` support, and mapping of `stdout`
- Language level `/` and `%` support for fast division using RP2040`s hardware dividers
- The function `runtime_init()` which performs minimal hardware initialisation (e.g. default PLL and clock configuration), and calls functions with `constructor` attributes before entering `main()`

`pico_standard_link` encapsulates the standard linker setup needed to configure the type of application binary layout in memory, and link to any additional C and/or C++ runtime libraries. It also includes the default `crt0`, which provides the initial entry point from the flash second stage bootloader, contains the initial vector table (later relocated to RAM), and initialises static data and RAM-resident code if the application is running from flash.

 NOTE

There is more high-level discussion of `pico_runtime` in Section 2.7

 TIP

Both `pico_runtime` and `pico_standard_link` are included with `pico_stdlib`

2.3.3. Hardware Support Libraries

These are individual libraries (`hardware_xxx`) providing actual APIs for interacting with each piece of physical hardware/peripheral. They are lightweight and provide only thin abstractions. The APIs are listed in Hardware APIs.

These libraries generally provide functions for configuring or interacting with the peripheral at a functional level, rather than accessing registers directly, e.g.

```
pio_sm_set_wrap(pio, sm, bottom, top);
```

rather than:

```
pio->sm[sm].execctrl =
    (pio->sm[sm].execctrl & ~(PIO_SM0_EXECCTRL_WRAP_TOP_BITS |
PIO_SM0_EXECCTRL_WRAP_BOTTOM_BITS)) |
    (bottom << PIO_SM0_EXECCTRL_WRAP_BOTTOM_LSB) |
    (top << PIO_SM0_EXECCTRL_WRAP_TOP_LSB);
```

The `hardware_` libraries are intended to have a very minimal runtime cost. They generally do not require any or much RAM, and do not rely on other runtime infrastructure. In general their only dependencies are the `hardware_structs` and `hardware_regs` libraries that contain definitions of memory-mapped register layout on RP2040. As such they can be used by low-level or other specialized applications that don't want to use the rest of the SDK libraries and runtime.

> **ℹ NOTE**
>
> `void pio_sm_set_wrap(PIO pio, uint sm, uint bottom, uint top) {}` is actually implemented as a `static inline` function in https://github.com/raspberrypi/pico-sdk/blob/master/src/rp2_common/hardware_pio/include/hardware/pio.h directly as shown above.
>
> Using `static inline` functions is common in SDK header files because such methods are often called with parameters that have fixed known values at compile time. In such cases, the compiler is often able to fold the code down to a single register write (or in this case a read, AND with a constant value, OR with a constant value, and a write) with no function call overhead. This tends to produce much smaller and faster binaries.

2.3.3.1. Hardware Claiming

The hardware layer does provide one small abstraction which is the notion of claiming a piece of hardware. This minimal system allows registration of peripherals or parts of peripherals (e.g. DMA channels) that are in use, and the ability to atomically claim free ones at runtime. The common use of this system - in addition to allowing for safe runtime allocation of resources - provides a better runtime experience for catching software misconfigurations or accidental use of the same piece hardware by multiple independent libraries that would otherwise be very painful to debug.

2.3.4. Hardware Structs Library

The `hardware_structs` library provides a set of C structures which represent the memory mapped layout of RP2040 registers in the system address space. This allows you to replace something like the following (which you'd write in C with the defines from the lower-level `hardware_regs`)

```
*(volatile uint32_t *)(PIO0_BASE + PIO_SM1_SHIFTCTRL_OFFSET) |=
PIO_SM1_SHIFTCTRL_AUTOPULL_BITS;
```

with something like this (where `pio0` is a pointer to type `pio_hw_t` at address PIO0_BASE):

```
pio0->sm[1].shiftctrl |= PIO_SM1_SHIFTCTRL_AUTOPULL_BITS;
```

The structures and associated pointers to memory mapped register blocks hide the complexity and potential error-prone-ness of dealing with individual memory locations, pointer casts and volatile access. As a bonus, the structs tend to produce better code with older compilers, as they encourage the reuse of a base pointer with offset load/stores, instead of producing a 32 bit literal for every register accessed.

The struct headers are named consistently with both the `hardware` libraries and the `hardware_regs` register headers. For example, if you access the `hardware_pio` library's functionality through `hardware/pio.h`, the `hardware_structs` library (a dependee of `hardware_pio`) contains a header you can include as `hardware/structs/pio.h` if you need to access a register directly, and this itself will pull in `hardware/regs/pio.h` for register field definitions. The PIO header is a bit lengthy to include here. `hardware/structs/pll.h` is a shorter example to give a feel for what these headers actually contain:

SDK: https://github.com/raspberrypi/pico-sdk/blob/master/src/rp2040/hardware_structs/include/hardware/structs/pll.h Lines 24 - 53

```
24  typedef struct {
25      _REG_(PLL_CS_OFFSET) // PLL_CS
26      // Control and Status
27      // 0x80000000 [31]    : LOCK (0): PLL is locked
28      // 0x00000100 [8]     : BYPASS (0): Passes the reference clock to the output instead of
    the divided VCO
```

```
29        // 0x0000003f [5:0]   : REFDIV (1): Divides the PLL input reference clock
30        io_rw_32 cs;
31
32        _REG_(PLL_PWR_OFFSET) // PLL_PWR
33        // Controls the PLL power modes
34        // 0x00000020 [5]     : VCOPD (1): PLL VCO powerdown
35        // 0x00000008 [3]     : POSTDIVPD (1): PLL post divider powerdown
36        // 0x00000004 [2]     : DSMPD (1): PLL DSM powerdown
37        // 0x00000001 [0]     : PD (1): PLL powerdown
38        io_rw_32 pwr;
39
40        _REG_(PLL_FBDIV_INT_OFFSET) // PLL_FBDIV_INT
41        // Feedback divisor
42        // 0x00000fff [11:0]  : FBDIV_INT (0): see ctrl reg description for constraints
43        io_rw_32 fbdiv_int;
44
45        _REG_(PLL_PRIM_OFFSET) // PLL_PRIM
46        // Controls the PLL post dividers for the primary output
47        // 0x00070000 [18:16] : POSTDIV1 (0x7): divide by 1-7
48        // 0x00007000 [14:12] : POSTDIV2 (0x7): divide by 1-7
49        io_rw_32 prim;
50 } pll_hw_t;
51
52 #define pll_sys_hw ((pll_hw_t *)PLL_SYS_BASE)
53 #define pll_usb_hw ((pll_hw_t *)PLL_USB_BASE)
```

The structure contains the layout of the hardware registers in a block, and some defines bind that layout to the base addresses of the *instances* of that peripheral in the RP2040 global address map.

Additionally, you can use one of the atomic `set`, `clear`, or `xor` address aliases of a piece of hardware to *set*, *clear* or *toggle* respectively the specified bits in a hardware register (as opposed to having the CPU perform a read/modify/write); e.g:

```
hw_set_alias(pio0)->sm[1].shiftctrl = PIO_SM1_SHIFTCTRL_AUTOPULL_BITS;
```

Or, equivalently

```
hw_set_bits(&pio0->sm[1].shiftctrl, PIO_SM1_SHIFTCTRL_AUTOPULL_BITS);
```

> **NOTE**
>
> The hardware atomic set/clear/XOR IO aliases are used extensively in the SDK libraries, to avoid certain classes of data race when two cores, or an IRQ and foreground code, are accessing registers concurrently.

> **NOTE**
>
> On RP2040 the atomic register aliases are a native part of the *peripheral*, not a CPU function, so the system DMA can also perform atomic set/clear/XOR operation on registers.

2.3.5. Hardware Registers Library

The `hardware_regs` library is a complete set of include files for all RP2040 registers, autogenerated from the hardware itself. This is all you need if you want to peek or poke a memory mapped register directly, however higher level libraries provide more user friendly ways of achieving what you want in C/C++.

For example, here is a snippet from `hardware/regs/sio.h`:

```c
// Description    : Single-cycle IO block
//                  Provides core-local and inter-core hardware for the two
//                  processors, with single-cycle access.
// =============================================================================
#ifndef HARDWARE_REGS_SIO_DEFINED
#define HARDWARE_REGS_SIO_DEFINED
// =============================================================================
// Register   : SIO_CPUID
// Description : Processor core identifier
//               Value is 0 when read from processor core 0, and 1 when read
//               from processor core 1.
#define SIO_CPUID_OFFSET 0x00000000
#define SIO_CPUID_BITS   0xffffffff
#define SIO_CPUID_RESET  "-"
#define SIO_CPUID_MSB    31
#define SIO_CPUID_LSB    0
#define SIO_CPUID_ACCESS "RO"
```

These header files are fairly heavily commented (the same information as is present in the datasheet register listings, or the SVD files). They define the offset of every register, and the layout of the fields in those registers, as well as the access type of the field, e.g. "RO" for read-only.

 TIP

> The headers in `hardware_regs` contain *only* comments and `#define` statements. This means they can be included from assembly files (`.S`, so the C preprocessor can be used), as well as C and C++ files.

2.3.6. TinyUSB Port

In addition to the core SDK libraries, we provide a RP2040 port of TinyUSB as the standard device and host USB support library within the SDK, and the SDK contains some build infrastructure for easily pulling this into your application.

The `tinyusb_dev` or `tinyusb_host` libraries within the SDK can be included in your application dependencies in `CMakeLists.txt` to add device or host support to your application respectively. Additionally, the `tinyusb_board` library is available to provide the additional "board support" code often used by TinyUSB demos. See the README in Pico Examples for more information and example code for setting up a fully functional application.

 IMPORTANT

> RP2040 USB hardware supports both Host and Device modes, but the two can not be used concurrently.

2.4. Directory Structure

We have discussed libraries such as `pico_stdlib` and `hardware_gpio` above. Imagine you wanted to add some code using RP2040's DMA controller to the `hello_world` example in `pico-examples`. To do this you need to add a dependency on another library, `hardware_dma`, which is not included by default by `pico_stdlib` (unlike, say, `hardware_uart`).

You would change your `CMakeLists.txt` to list both `pico_stdlib` and `hardware_dma` as dependencies of the `hello_world` target (executable). (Note the line breaks are not required)

```
target_link_libraries(hello_world
    pico_stdlib
    hardware_dma
)
```

And in your source code you would include the DMA hardware library header as such:

```
#include "hardware/dma.h"
```

Trying to include this header *without* listing `hardware_dma` as a dependency will fail, and this is due to how SDK files are organised into logical functional units on disk, to make it easier to add functionality in the future.

As an aside, this correspondence of `hardware_dma` → `hardware/dma.h` is the convention for *all* toplevel SDK library headers. The library is called `foo_bar` and the associated header is `foo/bar.h`. Some functions may be provided inline in the headers, others may be compiled and linked from additional `.c` files belonging to the library. Both of these require the relevant `hardware_` library to be listed as a dependency, either directly or through some higher-level bundle like `pico_stdlib`.

 NOTE

> Some libraries have additional headers which are located in `foo/bar/other.h`

You may want to actually find the files in question (although most IDEs will do this for you). The on disk files are actually split into multiple top-level directories. This is described in the next section.

2.4.1. Locations of Files

Whilst you may be focused on building a binary to run specifically on Raspberry Pi Pico, which uses a RP2040, the SDK is structured in a more general way. This is for two reasons:

1. To support other future chips in the RP2 family
2. To support testing of your code off device (this is *host* mode)

The latter is useful for writing and running unit tests, but also as you develop your software, for example your debugging code or work in progress software might actually be too big or use too much RAM to fit on the device, and much of the software complexity may be non-hardware-specific.

The code is thus split into top level directories as follows:

Table 1. Top-level directories

Path	Description
src/rp2040/	This contains the `hardware_regs` and `hardware_structs` libraries mentioned earlier, which are specific to RP2040.
src/rp2_common/	This contains the `hardware_` library implementations for individual hardware components, and `pico_` libraries or library implementations that are closely tied to RP2040 hardware. This is separate from `/src/rp2040` as there may be future revisions of RP2040, or other chips in the *RP2* family, which can use a common SDK and API whilst potentially having subtly different register definitions.
src/common/	This is code that is common to all builds. This is generally headers providing hardware abstractions for functionality which are simulated in host mode, along with a lot of the `pico_` library implementations which, to the extent they use hardware, do so only through the `hardware_` abstractions.

Path	Description
src/host/	This is a basic set of replacement SDK library implementations sufficient to get simple Raspberry Pi Pico applications running on your computer (Raspberry Pi OS, Linux, macOS or Windows using Cygwin or Windows Subsystem for Linux). This is not intended to be a fully functional simulator, however it is possible to inject additional implementations of libraries to provide more complete functionality.

There is a CMake variable `PICO_PLATFORM` that controls the environment you are building for:

When doing a regular RP2040 build (`PICO_PLATFORM=rp2040`, the default), you get code from `common`, `rp2_common` and `rp2040`; when doing a host build (`PICO_PLATFROM=host`), you get code from `common` and `host`.

Within each top-level directory, the libraries have the following structure (reading `foo_bar` as something like `hardware_uart` or `pico_time`):

```
top-level_dir/
top-level_dir/foo_bar/include/foo/bar.h    # header file
top-level_dir/foo_bar/CMakeLists.txt       # build configuration
top-level_dir/foo_bar/bar.c                # source file(s)
```

As a concrete example, we can list the `hardware_uart` directory under `pico-sdk/rp2_common` (you may also recall the `hardware_gpio` library we looked at earlier):

```
hardware_uart
├── CMakeLists.txt
├── include
│   └── hardware
│       └── uart.h
└── uart.c
```

`uart.h` contains function declarations and preprocessor defines for the `hardware_uart` library, as well as some inline functions that are expected to be particularly amenable to constant folding by the compiler. `uart.c` contains the implementations of more complex functions, such as calculating and setting up the divisors for a given UART baud rate.

 NOTE

> The directory `top-level_dir/foo_bar/include` is added as an include directory to the INTERFACE library `foo_bar`, which is what allows you to include `"foo/bar.h"` in your application

2.5. Conventions for Library Functions

This section covers some common patterns you will see throughout the SDK libraries, such as conventions for function names, how errors are reported, and the approach used to efficiently configure hardware with many register fields without having unreadable numbers of function arguments.

2.5.1. Function Naming Conventions

SDK functions follow a common naming convention for consistency and to avoid name conflicts. Some names are quite long, but that is deliberate to be as specific as possible about functionality, and of course because the SDK API is a C API and does not support function overloading.

2.5.1.1. Name prefix

Functions are prefixed by the library/functional area they belong to; e.g. public functions in the `hardware_dma` library are prefixed with `dma_`. Sometime the prefix refers to a sub group of library functionality (e.g. `channel_config_`)

2.5.1.2. Verb

A verb typically follows the prefix specifying that action performed by the function. `set_` and `get_` (or `is_` for booleans) are probably the most common and should always be present; i.e. a hypothetical method would be `oven_get_temperature()` and `food_add_salt()`, rather than `oven_temperature()` and `food_salt()`.

2.5.1.3. Suffixes

2.5.1.3.1. Blocking/Non-Blocking Functions and Timeouts

Table 2. SDK Suffixes for (non-)blocking functions and timeouts.

Suffix	Param	Description
(none)		The method is non-blocking, i.e. it does not wait on any external condition that could potentially take a long time.
`_blocking`		The method is blocking, and may potentially block indefinitely until some specific condition is met.
`_blocking_until`	`absolute_time_t until`	The method is blocking until some specific condition is met, however it will return early with a timeout condition (see Section 2.5.2) if the `until` time is reached.
`_timeout_ms`	`uint32_t timeout_ms`	The method is blocking until some specific condition is met, however it will return early with a timeout condition (see Section 2.5.2) after the specified number of milliseconds
`_timeout_us`	`uint64_t timeout_us`	The method is blocking until some specific condition is met, however it will return early with a timeout condition (see Section 2.5.2) after the specified number of microseconds

2.5.2. Return Codes and Error Handling

As mentioned earlier, there is a decision to be made as to whether/which functions return error codes that can be handled by the caller, and indeed whether the caller is likely to actually do something in response in an embedded environment. Also note that very often return codes are there to handle parameter checking, e.g. when asked to do something with the 27th DMA channel (when there are actually only 12).

In many cases checking for obviously invalid (likely program bug) parameters in (often inline) functions is prohibitively expensive in speed and code size terms, and therefore we need to be able to configure it on/off, which precludes return codes being returned for these exceptional cases.

The SDK follows two strategies:

1. Methods that can legitimately fail at runtime due to runtime conditions e.g. timeouts, dynamically allocated resource, can return a status which is either a `bool` indicating success or not, or an integer return code from the `PICO_ERROR_` family; non-error returns are `>= 0`.

2. Other items like invalid parameters, or failure to allocate resources which are deemed program bugs (e.g. two libraries trying to use the same statically assigned piece of hardware) do not affect a return code (usually the functions return `void`) and must cause some sort of exceptional event.

   As of right now the exceptional event is a C `assert`, so these checks are always disabled in release builds by default. Additionally most of the calls to `assert` are disabled by default for code/size performance (even in debug builds); You can set `PARAM_ASSERTIONS_ENABLE_ALL=1` or `PARAM_ASSERTIONS_DISABLE_ALL=1` in your build to change the default across the entire SDK, or say `PARAM_ASSERTIONS_ENABLED_I2C=0/1` to explicitly specify the behavior for the `hardware_i2c` module

In the future we expect to support calling a custom function to throw an exception in C++ or other environments where stack unwinding is possible.

3. Obviously sometimes the calling code whether it be user code or another higher level function, may not want the called function to assert on bad input, in which case it is the responsibility of the caller to check the validity (there are a good number of API functions provided that help with this) of their arguments, and the caller can then choose to provide a more flexible runtime error experience.

4. Finally some code may choose to "panic" directly if it detects an invalid state. A "panic" involves writing a message to standard output and then halting (by executing a breakpoint instruction). Panicking is a good response when it is undesirable to even attempt to continue given the current situation.

2.5.3. Use of Inline Functions

SDK libraries often contain a mixture of `static inline` functions in header files, and non-static functions in C source files. In particular, the `hardware_` libraries are likely to contain a higher proportion of inline function definitions in their headers. This is done for speed and code size.

The code space needed to setup parameters for a regular call to a small function in another compilation unit can be substantially larger than the function implementation. Compilers have their own metrics to decide when to inline function implementations at their call sites, but the use of `static inline` definitions gives the compiler more freedom to do this.

One reason this is *particularly* effective in the context of hardware register access is that these functions often:

1. Have relatively many parameters, which
2. Are immediately shifted and masked to combine with some register value, and
3. Are often constants known at compile time

So if the implementation of a hardware access function is inlined, the compiler can propagate the constant parameters through whatever bit manipulation and arithmetic that function may do, collapsing a complex function down to "please write this constant value to this constant address". Again, we are not forcing the compiler to do this, but the SDK consistently tries to give it freedom to do so.

The result is that there is generally no overhead using the lower-level `hardware_` functions as compared with using preprocessor macros with the `hardware_regs` definitions, and they tend to be much less error-prone.

2.5.4. Builder Pattern for Hardware Configuration APIs

The SDK uses a *builder pattern* for the more complex configurations, which provides the following benefits:

1. Readability of code (avoid "death by parameters" where a configuration function takes a dozen integers and booleans)
2. Tiny runtime code (thanks to the compiler)
3. Less brittle (the addition of another item to a hardware configuration will not break existing code)

Take the following hypothetical code example to (quite extensively) configure a DMA channel:

```
int dma_channel = 3;
dma_channel_config config = dma_get_default_channel_config(dma_channel);
channel_config_set_read_increment(&config, true);
channel_config_set_write_increment(&config, true);
```

```
channel_config_set_dreq(&config, DREQ_SPI0_RX);
channel_config_set_transfer_data_size(&config, DMA_SIZE_8);
dma_set_config(dma_channel, &config, false);
```

The value of `dma_channel` is known at compile time, so the compiler can replace `dma_channel` with 3 when generating code (*constant folding*). The `dma_` methods are `static inline` methods (from https://github.com/raspberrypi/pico-sdk/blob/master/src/rp2_common/hardware_dma/include/hardware/dma.h) meaning the implementations can be folded into your code by the compiler and, consequently, your constant parameters (like `DREQ_SPI0_RX`) are propagated though this local copy of the function implementation. The resulting code is usually smaller, and certainly faster, than the register shuffling caused by setting up a function call.

The net effect is that the compiler actually reduces all of the above to the following code:

Effective code produced by the C compiler for the DMA configuration

```
*(volatile uint32_t *)(DMA_BASE + DMA_CH3_AL1_CTRL_OFFSET) = 0x00089831;
```

It may seem counterintuitive that building up the configuration by passing a `struct` around, and committing the final result to the IO register, would be so much more compact than a series of direct register modifications using register field accessors. This is because the compiler is customarily forbidden from eliminating IO accesses (illustrated here with a `volatile` keyword), with good reason. Consequently it's easy to unwittingly generate code that repeatedly puts a value into a register and pulls it back out again, changing a few bits at a time, when we only care about the final value of the register. The configuration pattern shown here avoids this common pitfall.

 NOTE

> The SDK code is designed to make builder patterns efficient in both *Release* and *Debug* builds. Additionally, even if not all values are known constant at compile time, the compiler can still produce the most efficient code possible based on the values that are known.

2.6. Customisation and Configuration Using Preprocessor variables

The SDK allows use of compile time definitions to customize the behavior/capabilities of libraries, and to specify settings (e.g. physical pins) that are unlikely to be changed at runtime This allows for much smaller more efficient code, and avoids additional runtime overheads and the inclusion of code for configurations you *might* choose at runtime even though you actually don't (e.g. support PWM audio when you are only using I2S)!

Remember that because of the use of *INTERFACE* libraries, all the libraries your application(s) depend on are built from source for each application in your build, so you can even build multiple variants of the same application with different baked in behaviors.

Appendix B has a comprehensive list of the available preprocessor defines, what they do, and what their default values are.

Preprocessor variables may be specified in a number of ways, described in the following sections.

> **ℹ NOTE**
>
> Whether compile time configuration or runtime configuration or both is supported/required is dependent on the particular library itself. The general philosophy however, is to allow sensible default behavior without the user specifying any settings (beyond those provided by the board configuration).

2.6.1. Preprocessor Variables via Board Configuration File

Many of the common configuration settings are actually related to the particular RP2040 board being used, and include default pin settings for various SDK libraries. The board being used is specified via the `PICO_BOARD` CMake variable which may be specified on the CMake command line or in the environment. The default `PICO_BOARD` if not specified is `pico`.

The board configuration provides a header file which specifies defaults if not otherwise specified; for example https://github.com/raspberrypi/pico-sdk/blob/master/src/boards/include/boards/pico.h specifies

```
#ifndef PICO_DEFAULT_LED_PIN
#define PICO_DEFAULT_LED_PIN 25
#endif
```

The header `my_board_name.h` is included by all other SDK headers as a result of setting `PICO_BOARD=my_board_name`. You may wish to specify your own board configuration in which case you can set PICO_BOARD_HEADER_DIRS in the environment or CMake to a semicolon separated list of paths to search for `my_board_name.h`.

2.6.2. Preprocessor Variables Per Binary or Library via CMake

We could modify the https://github.com/raspberrypi/pico-examples/blob/master/hello_world/CMakeLists.txt with `target_compile_definitions` to specify an alternate set of UART pins to use.

Modified hello_world CMakeLists.txt specifying different UART pins

```
add_executable(hello_world
    hello_world.c
)

# SPECIFY two preprocessor definitions for the target hello_world
target_compile_definitions(hello_world PRIVATE
    PICO_DEFAULT_UART_TX_PIN=16
    PICO_DEFAULT_UART_RX_PIN=17
)

# Pull in our pico_stdlib which aggregates commonly used features
target_link_libraries(hello_world pico_stdlib)

# create map/bin/hex/uf2 file etc.
pico_add_extra_outputs(hello_world)
```

The `target_compile_definitions` specifies preprocessor definitions that will be passed to the compiler for every source file in the target `hello_world` (which as mentioned before includes *all* of the sources for all dependent *INTERFACE* libraries). `PRIVATE` is required by CMake to specify the scope for the compile definitions. Note that all preprocessor definitions used by the SDK have a `PICO_` prefix.

2.7. SDK Runtime

For those coming from non-embedded programming, or from other devices, this section will give you an idea of how various C/C++ language level concepts are handled within the SDK

2.7.1. Standard Input/Output (stdio) Support

The SDK runtime packages a lightweight `printf` library by Marco Paland, linked as `pico_printf`. It also contains infrastructure for routing `stdout` and `stdin` to various hardware interfaces, which is documented under `pico_stdio`:

- A UART interface specified by a board configuration header. The default for Raspberry Pi Pico is 115200 baud on GPIO0 (TX) and GPIO1 (RX)
- A USB CDC ACM virtual serial port, using TinyUSB's CDC support. The virtual serial device can be accessed through RP2040's dedicated USB hardware interface, in Device mode.
- (Experimental) minimal semihosting support to direct `stdout` to an external debug host connected via the Serial Wire Debug link on RP2040

These can be accessed using standard calls like `printf`, `puts`, `getchar`, found in the standard `<stdio.h>` header. By default, `stdout` converts bare linefeed characters to carriage return plus linefeed, for better display in a terminal emulator. This can be disabled at runtime, at build time, or the CR-LF support can be completely removed.

`stdout` is broadcast to all interfaces that are enabled, and `stdin` is collected from all interfaces which are enabled and support input. Since some of the interfaces, particularly USB, have heavy runtime and binary size cost, only the UART interface is included by default. You can add/remove interfaces for a given program at build time with e.g.

```
pico_enable_stdio_usb(target_name, 1)
```

2.7.2. Floating-point Support

The SDK provides a highly optimized single and double precision floating point implementation. In addition to being fast, many of the functions are actually implemented using support provided in the RP2040 bootrom. This means the interface from your code to the ROM floating point library has very minimal impact on your program size, certainly using dramatically less flash storage than including the standard floating point routines shipped with your compiler.

The physical ROM storage on RP2040 has single-cycle access (with a dedicated arbiter on the RP2040 busfabric), and accessing code stored here does not put pressure on the flash cache or take up space in memory, so not only are the routines fast, the rest of your code will run faster due them being resident in ROM.

This implementation is used by default as it is the best choice in the majority of cases, however it is also possible to switch to using the regular compiler soft floating point support.

2.7.2.1. Functions

The SDK provides implementations for all the standard functions from `math.h`. Additional functions can be found in `pico/float.h` and `pico/double.h`.

2.7.2.2. Speed/Tradeoffs

The overall goal for the bootrom floating-point routines is to achieve good performance within a small footprint, the emphasis being more on improved performance for the basic operations (add, subtract, multiply, divide and square root, and all conversion functions), and more on reduced footprint for the scientific functions (trigonometric functions, logarithms and exponentials).

The IEEE single- and double-precision data formats are used throughout, but in the interests of reducing code size, input denormals are treated as zero and output denormals are flushed to zero, and output NaNs are rendered as infinities. Only the round-to-nearest, even-on-tie rounding mode is supported. Traps are not supported. Whether input NaNs are treated as infinities or propagated is configurable.

The five basic operations (add, subtract, multiply, divide, sqrt) return results that are always correctly rounded (round-to-nearest).

The scientific functions always return results within 1 ULP (unit in last place) of the exact result. In many cases results are better.

The scientific functions are calculated using internal fixed-point representations so accuracy (as measured in ULP error rather than in absolute terms) is poorer in situations where converting the result back to floating point entails a large normalising shift. This occurs, for example, when calculating the sine of a value near a multiple of pi, the cosine of a value near an odd multiple of pi/2, or the logarithm of a value near 1. Accuracy of the tangent function is also poorer when the result is very large. Although covering these cases is possible, it would add considerably to the code footprint, and there are few types of program where accuracy in these situations is essential.

The following table shows the results from a benchmark

> **NOTE**
>
> Whilst the SDK floating point support makes use of the routines in the RP2040 bootrom, it hides some of the limitations of the raw ROM functions (e.g. limited sin/cos range), in order to be largely indistinguishable from the compiler-provided functionality. Certain smaller functions have also been re-implemented for even more speed outside of the limited bootrom space.

Table 3. SDK implementation vs GCC 9 implementation for ARM AEABI floating point functions (these unusually named functions provide the support for basic operations on float and double types)

Function	ROM/SDK (μs)	GCC 9 (μs)	Performance Ratio
__aeabi_fadd	72.4	99.8	138%
__aeabi_fsub	86.7	133.6	154%
__aeabi_frsub	89.8	140.6	157%
__aeabi_fmul	61.5	145	236%
__aeabi_fdiv	74.7	437.5	586%
__aeabi_fcmplt	39	61.1	157%
__aeabi_fcmple	40.5	61.1	151%
__aeabi_fcmpgt	40.5	61.2	151%
__aeabi_fcmpge	41	61.2	149%
__aeabi_fcmpeq	40	41.5	104%
__aeabi_dadd	99.4	142.5	143%
__aeabi_dsub	114.2	182	159%
__aeabi_drsub	108	181.2	168%
__aeabi_dmul	168.2	338	201%
__aeabi_ddiv	197.1	412.2	209%
__aeabi_dcmplt	53	88.3	167%
__aeabi_dcmple	54.6	88.3	162%
__aeabi_dcmpgt	54.4	86.6	159%
__aeabi_dcmpge	55	86.6	157%

__aeabi_dcmpeq	54	64.3	119%
__aeabi_f2iz	17	24.5	144%
__aeabi_f2uiz	42.5	106.5	251%
__aeabi_f2lz	63.1	1240.5	1966%
__aeabi_f2ulz	46.1	1157	2510%
__aeabi_i2f	43.5	63	145%
__aeabi_ui2f	41.5	55.8	134%
__aeabi_l2f	75.2	643.3	855%
__aeabi_ul2f	71.4	531.5	744%
__aeabi_d2iz	30.6	44.1	144%
__aeabi_d2uiz	75.7	159.1	210%
__aeabi_d2lz	81.2	1267.8	1561%
__aeabi_d2ulz	65.2	1148.3	1761%
__aeabi_i2d	44.4	61.9	139%
__aeabi_ui2d	43.4	51.3	118%
__aeabi_l2d	104.2	559.3	537%
__aeabi_ul2d	102.2	458.1	448%
__aeabi_f2d	20	31	155%
__aeabi_d2f	36.4	66	181%

2.7.2.3. Configuration and Alternate Implementations

There are three different floating point implementations provided

Name	Description
default	The default; equivalent to pico
pico	Use the fast/compact SDK/bootrom implementations
compiler	Use the standard compiler provided soft floating point implementations
none	Map all functions to a runtime assertion. You can use this when you know you don't want any floating point support to make sure it isn't accidentally pulled in by some library.

These settings can be set independently for both "float" and "double":

For "float" you can call `pico_set_float_implementation(TARGET NAME)` in your `CMakeLists.txt` to choose a specific implementation for a particular target, or set the CMake variable `PICO_DEFAULT_FLOAT_IMPL` to `pico_float_NAME` to set the default.

For "double" you can call `pico_set_double_implementation(TARGET NAME)` in your `CMakeLists.txt` to choose a specific implementation for a particular target, or set the CMake variable `PICO_DEFAULT_DOUBLE_IMPL` to `pico_double_NAME` to set the default.

> 💡 **TIP**
>
> The `pico` floating point library adds very little to your binary size, however it must include implementations for any used functions that are not present in V1 of the bootrom, which is present on early Raspberry Pi Pico boards. If you know that you are only using RP2040s with V2 of the bootrom, then you can specify defines `PICO_FLOAT_SUPPORT_ROM_V1=0` and `PICO_DOUBLE_SUPPORT_ROM_V1=0` so the extra code will not be included. Any use of those functions on a RP2040 with a V1 bootrom will cause a panic at runtime. See the **RP2040 Datasheet** for more specific details of the bootrom functions.

2.7.2.3.1. NaN Propagation

The SDK implementation by default treats input *NaNs* as infinites. If you require propagation of NaN inputs to outputs and NaN outputs for domain errors, then you can set the compile definitions `PICO_FLOAT_PROPAGATE_NANS` and `PICO_DOUBLE_PROPAGATE_NANS` to 1, at the cost of a small runtime overhead.

2.7.3. Hardware Divider

The SDK includes optimized 32- and 64-bit division functions accelerated by the RP2040 hardware divider, which are seamlessly integrated with the C `/` and `%` operators. The SDK also supplies a high level API which includes combined quotient and remainder functions for 32- and 64-bit, also accelerated by the hardware divider.

See Figure 1 and Figure 2 for 32-bit and 64-bit integer divider comparison.

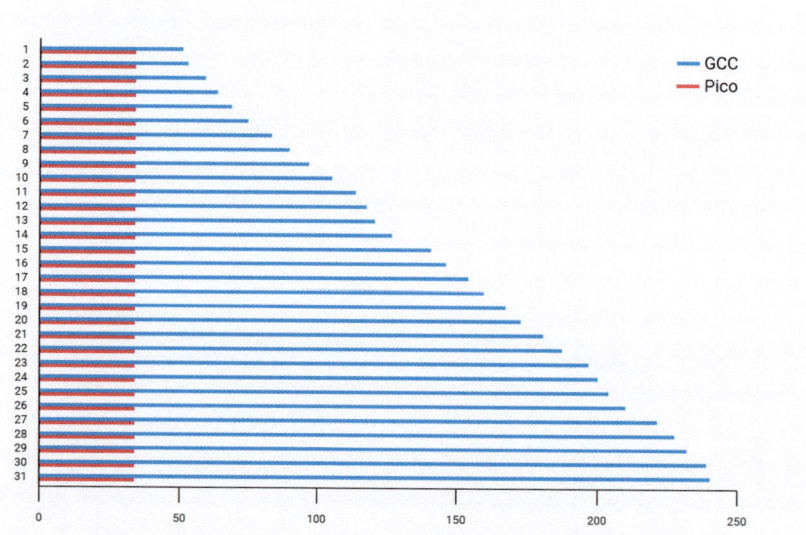

Figure 1. 32-bit divides by divider size using GCC library (blue), or the SDK library (red) with the RP2040 hardware divider.

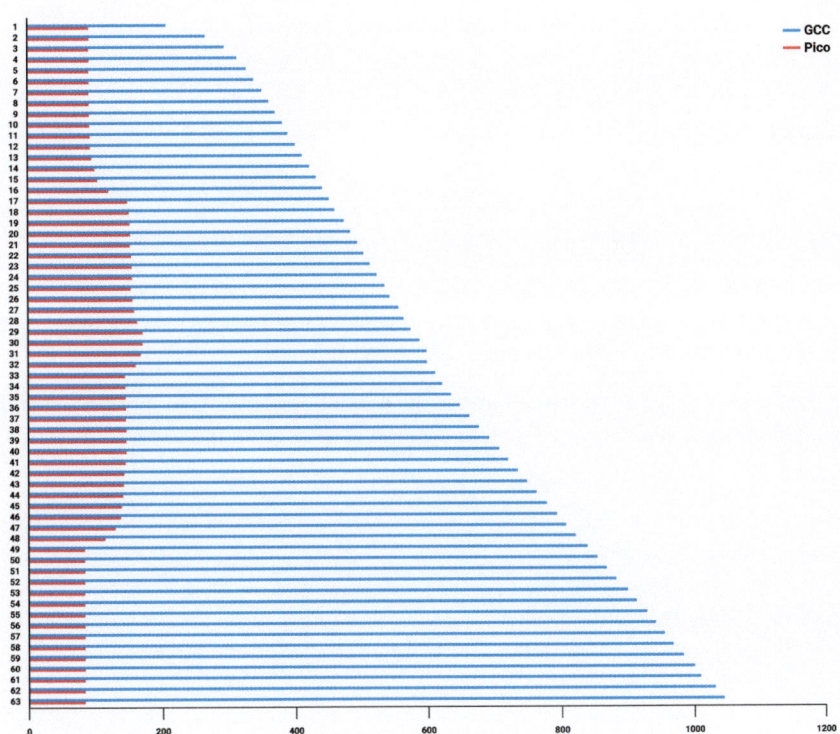

Figure 2. *64-bit divides by divider size using GCC library (blue), or the SDK library (red) with the RP2040 hardware divider.*

2.8. Multi-core support

Multi-core support should be familiar to those used to programming with threads in other environments. The second core is just treated as a second *thread* within your application; initially the second core (`core1` as it is usually referred to; the main application thread runs on `core0`) is halted, however you can start it executing some function in parallel from your main application thread.

Core 1 (the second core) is started by calling `multicore_launch_core1(some_function_pointer);` on core 0, which wakes the core from its low-power sleep state and provides it with its entry point — some function you have provided which hopefully has a descriptive name like `void core1_main() { }`. This function, as well as others such as pushing and popping data through the inter-core mailbox FIFOs, is listed under `pico_multicore`.

Care should be taken with calling C library functions from both cores simultaneously as they are generally not designed to be thread safe. You can use the `mutex_` API provided by the SDK in the `pico_sync` library (https://github.com/raspberrypi/pico-sdk/blob/master/src/common/pico_sync/include/pico/mutex.h) from within your own code.

> **NOTE**
>
> That the SDK version of printf is always safe to call from both cores. `malloc`, `calloc` and `free` are additionally wrapped to make it thread safe when you include the `pico_multicore` as a convenience for C++ programming, where some object allocations may not be obvious.

2.9. Using C++

The SDK has a C style API, however the SDK headers may be safely included from C++ code, and the functions called (they are declared with C linkage).

C++ files are integrated into SDK projects in the same way as C files: listing them in your `CMakeLists.txt` file under either the `add_executable()` entry, or a separate `target_sources()` entry to append them to your target.

To save space, exception handling is disabled by default; this can be overridden with the CMake environment variable
PICO_CXX_ENABLE_EXCEPTIONS=1. There are a handful of other C++ related `PICO_CXX` vars listed in Appendix C.

2.10. Next Steps

This has been quite a deep dive. If you've somehow made it through this chapter *without* building any software, now would be a perfect time to divert to the **Getting started with Raspberry Pi Pico** book, which has detailed instructions on connecting to your RP2040 board and loading an application built with the SDK.

Chapter 3 gives some background on RP2040's unique Programmable I/O subsystem, and walks through building some applications which use PIO to talk to external hardware.

Chapter 4 is a comprehensive listing of the SDK APIs. The APIs are listed according to groups of related functionality (e.g. low-level hardware access).

Chapter 3. Using programmable I/O (PIO)

3.1. What is Programmable I/O (PIO)?

Programmable I/O (PIO) is a new piece of hardware developed for RP2040. It allows you to create new types of (or additional) hardware interfaces on your RP2040-based device. If you've looked at fixed peripherals on a microcontroller, and thought "I want to add 4 more UARTs", or "I'd like to output DPI video", or even "I need to communicate with this cursed serial device I found on AliExpress, but no machine has hardware support", then you will have fun with this chapter.

PIO hardware is described extensively in chapter 3 of the RP2040 Datasheet. This is a companion to that text, focussing on how, when and why to use PIO in your software. To start, we're going to spend a while discussing why I/O is hard, what the current options are, and what PIO does differently, before diving into some software tutorials. We will also try to illuminate some of the more important parts of the hardware along the way, but will defer to the datasheet for full explanations.

 TIP

> You can skip to the first software tutorial if you'd prefer to dive straight in.

3.1.1. Background

Interfacing with other digital hardware components is hard. It often happens at very high frequencies (due to amounts of data that need to be transferred), and has very exact timing requirements.

3.1.2. I/O Using dedicated hardware on your PC

Traditionally, on your desktop or laptop computer, you have one option for hardware interfacing. Your computer has high speed USB ports, HDMI outputs, PCIe slots, SATA drive controllers etc. to take care of the tricky and time sensitive business of sending and receiving ones and zeros, and responding with minimal latency or interruption to the graphics card, hard drive etc. on the other end of the hardware interface.

The custom hardware components take care of specific tasks that the more general multi-tasking CPU is not designed for. The operating system drivers perform higher level management of what the hardware components do, and coordinate data transfers via DMA to/from memory from the controller and receive IRQs when high level tasks need attention. These interfaces are purpose-built, and if you have them, you should use them.

3.1.3. I/O Using dedicated hardware on your Raspberry Pi or microcontroller

Not so common on PCs: your Raspberry Pi or microcontroller is likely to have dedicated hardware on chip for managing UART, I2C, SPI, PWM, I2S, CAN bus and more over *general purpose I/O* pins (GPIOs). Like USB controllers (also found on some microcontrollers, including the RP2040 on Raspberry Pi Pico), I2C and SPI are general purpose buses which connect to a wide variety of external hardware, using the same piece of on-chip hardware. This includes sensors, external flash, EEPROM and SRAM memories, GPIO expanders, and more, all of them widely and cheaply available. Even HDMI uses I2C to communicate video timings between Source and Sink, and there is probably a microcontroller *embedded* in your TV to handle this.

These protocols are simpler to integrate into very low-cost *devices* (i.e. not the host), due to their relative simplicity and modest speed. This is important for chips with mostly analogue or high-power circuitry: the silicon fabrication techniques used for these chips do not lend themselves to high speed or gate count, so if your switchmode power supply controller has some serial configuration interface, it is likely to be something like I2C. The number of traces routed on the circuit board, the number of pins required on the device package, and the PCB technology required to maintain signal integrity are also factors in the choice of these protocols. A microcontroller needs to communicate with these devices to be part of a larger *embedded system*.

This is all very well, but the area taken up by these individual serial peripherals, and the associated cost, often leaves you with a limited menu. You may end up paying for a bunch of stuff you don't need, and find yourself without enough of what you really want. Of course you are out of luck if your microcontroller does not have dedicated hardware for the type of hardware device you want to attach (although in some cases you may be able to bridge over USB, I2C or SPI at the cost of buying external hardware).

3.1.4. I/O Using software control of GPIOs (*"bit-banging"*)

The third option on your Raspberry Pi or microcontroller — any system with GPIOs which the processor(s) can access easily — is to use the CPU to wiggle (and listen to) the GPIOs at dizzyingly high speeds, and hope to do so with sufficiently correct timing that the external hardware still understands the signals.

As a bit of background it is worth thinking about types of hardware that you might want to interface, and the approximate signalling speeds involved:

Table 4. Types of hardware

Interface Speed	Interface
1-10Hz	Push buttons, indicator LEDs
300Hz	HDMI CEC
10-100kHz	Temperature sensors (DHT11), one-wire serial
<100kHz	I2C Standard mode
22-100+kHz	PCM audio
300+kHz	PWM audio
400-1200kHz	WS2812 LED string
10-3000kHz	UART serial
12MHz	USB Full Speed
1-100MHz	SPI
20-300MHz	DPI/VGA video
480MHz	USB High Speed
10-4000MHz	Ethernet LAN
12-4000MHz	SD card
250-20000MHz	HDMI/DVI video

"Bit-Banging" (i.e. using the processor to hammer out the protocol via the GPIOs) is very hard. The processor isn't really designed for this. It has other work to do… for slower protocols you might be able to use an IRQ to wake up the processor from what it was doing fast enough (though latency here is a concern) to send the next bit(s). Indeed back in the early days of PC sound it was not uncommon to set a hardware timer interrupt at 11kHz and write out one 8-bit PCM sample every interrupt for some rather primitive sounding audio!

Doing that on a PC nowadays is laughed at, even though they are many order of magnitudes faster than they were back then. As processors have become faster in terms of overwhelming number-crunching brute force, the layers of software and hardware between the processor and the outside world have also grown in number and size. In response to the growing distance between processors and memory, PC-class processors keep many hundreds of instructions in-flight on a single core at once, which has drawbacks when trying to switch rapidly between hard real time tasks. However, IRQ-based bitbanging can be an effective strategy on simpler embedded systems.

Above certain speeds — say a factor of 1000 below the processor clock speed — IRQs become impractical, in part due to the timing uncertainty of actually *entering* an interrupt handler. The alternative when *"bit-banging"* is to sit the processor in a carefully timed loop, often painstakingly written in assembly, trying to make sure the GPIO reading and writing happens on the exact cycle required. This is really really hard work if indeed possible at all. Many heroic hours and likely thousands of GitHub repositories are dedicated to the task of doing such things (a large proportion of them for LED strings).

Additionally of course, your processor is now busy doing the *"bit-banging"*, and cannot be used for other tasks. If your processor is interrupted even for a few microseconds to attend to one of the hard peripherals it is also responsible for, this can be fatal to the timing of any bit-banged protocol. The greater the ratio between protocol speed and processor speed, the more cycles your processor will spend uselessly idling in between GPIO accesses. Whilst it is eminently possible to drive a 115200 baud UART output using only software, this has a cost of >10,000 cycles per byte if the processor is running at 133MHz, which may be poor investment of those cycles.

Whilst dealing with something like an LED string is possible using *"bit-banging"*, once your hardware protocol gets faster to the point that it is of similar order of magnitude to your system clock speed, there is really not much you can hope to do. The main case where software GPIO access is the *best* choice is LEDs and push buttons.

Therefore you're back to custom hardware for the protocols you know up front you are going to want (or more accurately, the chip designer thinks you might need).

3.1.5. Programmable I/O Hardware using FPGAs and CPLDs

A *field-programmable gate array* (FPGA), or its smaller cousin, the *complex programmable logic device* (CPLD), is in many ways the perfect solution for tailor-made I/O requirements, whether that entails an unusual type or unusual mixture of interfaces. FPGAs are chips with a configurable logic fabric — effectively a sea of gates and flipflops, some other special digital function blocks, and a routing fabric to connect them — which offer the same level of design flexibility available to chip designers. This brings with it all the advantages of dedicated I/O hardware:

- Absolute precision of protocol timing (within limitations of your clock source)
- Capable of very high I/O throughput
- Offload simple, repetitive calculations that are part of the I/O standard (checksums)
- Present a simpler interface to host software; abstract away details of the protocol, and handle these details internally.

The main drawback of FPGAs in embedded systems is their cost. They also present a very unfamiliar programming model to those well-versed in embedded software: you are not programming at all, but rather designing digital hardware. One you have your FPGA you will still need some other processing element in your system to run control software, unless you are using an FPGA expensive enough to either fit a soft CPU core, or contain a hardened CPU core alongside the FPGA fabric.

eFPGAs (embedded FPGAs) are available in some microcontrollers: a slice of FPGA logic fabric integrated into a more conventional microcontroller, usually with access to some GPIOs, and accessible over the system bus. These are attractive from a system integration point of view, but have a significant area overhead compared with the usual serial peripherals found on a microcontroller, so either increase the cost and power dissipation, or are very limited in size. The issue of programming complexity still remains in eFPGA-equipped systems.

3.1.6. Programmable I/O Hardware using PIO

The PIO subsystem on RP2040 allows you to write small, simple programs for what are called *PIO state machines*, of which RP2040 has eight split across two PIO *instances*. A state machine is responsible for setting and reading one or more GPIOs, buffering data to or from the processor (or RP2040's ultra-fast DMA subsystem), and notifying the processor, via IRQ or polling, when data or attention is needed.

These programs operate with cycle accuracy at up to system clock speed (or the program clocks can be divided down to run at slower speeds for less frisky protocols).

PIO state machines are much more compact than the general-purpose Cortex-M0+ processors on RP2040. In fact, they are similar in size (and therefore cost) to a standard SPI peripheral, such as the PL022 SPI also found on RP2040, because much of their area is spent on components which are common to all serial peripherals, like FIFOs, shift registers and clock dividers. The instruction set is small and regular, so not much silicon is spent on decoding the instructions. There is no need to feel guilty about dedicating a state machine solely to a single I/O task, since you have 8 of them!

In spite of this, a PIO state machine gets a lot *more* done in one cycle than a Cortex-M0+ when it comes to I/O: for example, sampling a GPIO value, toggling a clock signal and pushing to a FIFO all in one cycle, every cycle. The trade-off is that a PIO state machine is not remotely capable of running general purpose software. As we shall see though, programming a PIO state machine is quite familiar for anyone who has written assembly code before, and the small instruction set should be fairly quick to pick up for those who haven't.

For simple hardware protocols - such as PWM or duplex SPI - a single PIO state machine can handle the task of implementing the hardware interface all on its own. For more involved protocols such as SDIO or DPI video you may end up using two or three.

 TIP

> If you are ever tempted to *"bit-bang"* a protocol on RP2040, don't! Use the PIO instead. Frankly this is true for anything that repeatedly reads or writes from GPIOs, but certainly anything which aims to transfer data.

3.2. Getting started with PIO

It is possible to write PIO programs both within the C++ SDK and directly from MicroPython.

Additionally the future intent is to add APIs to trivially have new UARTs, PWM channels etc created for you, using a menu of pre-written PIO programs, but for now you'll have to follow along with example code and do that yourself.

3.2.1. A First PIO Application

Before getting into all of the fine details of the PIO assembly language, we should take the time to look at a small but complete application which:

1. Loads a program into a PIO's instruction memory
2. Sets up a PIO state machine to run the program
3. Interacts with the state machine once it is running.

The main ingredients in this recipe are:

- A PIO program
- Some software, written in C, to run the whole show
- A CMake file describing how these two are combined into a program image to load onto a RP2040-based development board

> **TIP**
>
> The code listings in this section are all part of a complete application on GitHub, which you can build and run. Just click the link above each listing to go to the source. In this section we are looking at the `pio/hello_pio` example in `pico-examples`. You might choose to build this application and run it, to see what it does, before reading through this section.

> **NOTE**
>
> The focus here is on the main moving parts required to use a PIO program, not so much on the PIO program itself. This is a lot to take in, so we will stay high-level in this example, and dig in deeper on the next one.

3.2.1.1. PIO Program

This is our first PIO program listing. It's written in PIO assembly language.

Pico Examples: https://github.com/raspberrypi/pico-examples/blob/master/pio/hello_pio/hello.pio *Lines 7 - 15*

```
 7  .program hello
 8
 9  ; Repeatedly get one word of data from the TX FIFO, stalling when the FIFO is
10  ; empty. Write the least significant bit to the OUT pin group.
11
12  loop:
13      pull
14      out pins, 1
15      jmp loop
```

The `pull` instruction takes one data item from the transmit FIFO buffer, and places it in the *output shift register* (OSR). Data moves from the FIFO to the OSR one word (32 bits) at a time. The OSR is able to *shift* this data out, one or more bits at a time, to further destinations, using an `out` instruction.

> **FIFOs?**
>
> FIFOs are data queues, implemented in hardware. Each state machine has two FIFOs, between the state machine and the system bus, for data travelling out of (TX) and into (RX) the chip. Their name (*first in, first out*) comes from the fact that data appears at the FIFO's output in the same order as it was presented to the FIFO's input.

The `out` instruction here takes one bit from the data we just `pull`-ed from the FIFO, and writes that data to some pins. We will see later how to decide which pins these are.

The `jmp` instruction jumps back to the `loop:` label, so that the program repeats indefinitely. So, to sum up the function of this program: repeatedly take one data item from a FIFO, take one bit from this data item, and write it to a pin.

Our `.pio` file also contains a helper function to set up a PIO state machine for correct execution of this program:

Pico Examples: https://github.com/raspberrypi/pico-examples/blob/master/pio/hello_pio/hello.pio *Lines 18 - 33*

```
18  static inline void hello_program_init(PIO pio, uint sm, uint offset, uint pin) {
19      pio_sm_config c = hello_program_get_default_config(offset);
20
21      // Map the state machine's OUT pin group to one pin, namely the `pin`
22      // parameter to this function.
23      sm_config_set_out_pins(&c, pin, 1);
```

```
24        // Set this pin's GPIO function (connect PIO to the pad)
25        pio_gpio_init(pio, pin);
26        // Set the pin direction to output at the PIO
27        pio_sm_set_consecutive_pindirs(pio, sm, pin, 1, true);
28
29        // Load our configuration, and jump to the start of the program
30        pio_sm_init(pio, sm, offset, &c);
31        // Set the state machine running
32        pio_sm_set_enabled(pio, sm, true);
33  }
```

Here the main thing to set up is the GPIO we intend to output our data to. There are three things to consider here:

1. The state machine needs to be told which GPIO or GPIOs to output to. There are four different pin groups which are used by different instructions in different situations; here we are using the out pin group, because we are just using an `out` instruction.

2. The *GPIO* also needs to be told that PIO is in control of it (GPIO function select)

3. If we are using the pin for output only, we need to make sure that PIO is driving the *output enable* line high. PIO can drive this line up and down programmatically using e.g. an `out pindirs` instruction, but here we are setting it up before starting the program.

3.2.1.2. C Program

PIO won't do anything until it's been configured properly, so we need some software to do that. The PIO file we just looked at — `hello.pio` — is converted automatically (we will see later how) into a header containing our assembled PIO program binary, any helper functions we included in the file, and some useful information about the program. We include this as `hello.pio.h`.

Pico Examples: https://github.com/raspberrypi/pico-examples/blob/master/pio/hello_pio/hello.c

```c
1  /**
2   * Copyright (c) 2020 Raspberry Pi (Trading) Ltd.
3   *
4   * SPDX-License-Identifier: BSD-3-Clause
5   */
6
7  #include "pico/stdlib.h"
8  #include "hardware/pio.h"
9  // Our assembled program:
10 #include "hello.pio.h"
11
12 int main() {
13 #ifndef PICO_DEFAULT_LED_PIN
14 #warning pio/hello_pio example requires a board with a regular LED
15 #else
16     // Choose which PIO instance to use (there are two instances)
17     PIO pio = pio0;
18
19     // Our assembled program needs to be loaded into this PIO's instruction
20     // memory. This SDK function will find a location (offset) in the
21     // instruction memory where there is enough space for our program. We need
22     // to remember this location!
23     uint offset = pio_add_program(pio, &hello_program);
24
25     // Find a free state machine on our chosen PIO (erroring if there are
26     // none). Configure it to run our program, and start it, using the
27     // helper function we included in our .pio file.
28     uint sm = pio_claim_unused_sm(pio, true);
29     hello_program_init(pio, sm, offset, PICO_DEFAULT_LED_PIN);
```

```
30
31      // The state machine is now running. Any value we push to its TX FIFO will
32      // appear on the LED pin.
33      while (true) {
34          // Blink
35          pio_sm_put_blocking(pio, sm, 1);
36          sleep_ms(500);
37          // Blonk
38          pio_sm_put_blocking(pio, sm, 0);
39          sleep_ms(500);
40      }
41 #endif
42 }
```

You might recall that RP2040 has two PIO blocks, each of them with four state machines. Each PIO block has a 32-slot instruction memory which is visible to the four state machines in the block. We need to load our program into this instruction memory before any of our state machines can run the program. The function `pio_add_program()` finds free space for our program in a given PIO's instruction memory, and loads it.

> **32 Instructions?**
>
> This may not sound like a lot, but the PIO instruction set can be *very* dense once you fully explore its features. A perfectly serviceable UART transmit program can be implemented in four instructions, as shown in the `pio/uart_tx` example in `pico-examples`. There are also a couple of ways for a state machine to execute instructions from other sources — like directly from the FIFOs — which you can read all about in the **RP2040 Datasheet**.

Once the program is loaded, we find a free state machine and tell it to run our program. There is nothing stopping us from ordering multiple state machines to run the same program. Likewise, we could instruct each state machine to run a *different* program, provided they all fit into the instruction memory at once.

We're configuring this state machine to output its data to the LED on your Raspberry Pi Pico board. If you have already built and run the program, you probably noticed this already!

At this point, the state machine is running autonomously. The state machine will immediately *stall*, because it is waiting for data in the TX FIFO, and we haven't provided any. The processor can push data directly into the state machine's TX FIFO using the `pio_sm_put_blocking()` function. (`_blocking` because this function stalls the processor when the TX FIFO is full.) Writing a 1 will turn the LED on, and writing a 0 will turn the LED off.

3.2.1.3. CMake File

We have two lovely text files sat on our computer, with names ending with `.pio` and `.c`, but they aren't doing us much good there. A CMake file describes how these are built into a binary suitable for loading onto your Raspberry Pi Pico or other RP2040-based board.

Pico Examples: https://github.com/raspberrypi/pico-examples/blob/master/pio/hello_pio/CMakeLists.txt

```
1 add_executable(hello_pio)
2
3 pico_generate_pio_header(hello_pio ${CMAKE_CURRENT_LIST_DIR}/hello.pio)
4
5 target_sources(hello_pio PRIVATE hello.c)
6
7 target_link_libraries(hello_pio PRIVATE
8         pico_stdlib
9         hardware_pio
10        )
```

```
11
12 pico_add_extra_outputs(hello_pio)
13
14 # add url via pico_set_program_url
15 example_auto_set_url(hello_pio)
```

- `add_executable()`: Declare that we are building a program called `hello_pio`

- `pico_generate_pio_header()`: Declare that we have a PIO program, `hello.pio`, which we want to be built into a C header for use with our program

- `target_sources()`: List the source code files for our `hello_pio` program. In this case, just one C file.

- `target_link_libraries()`: Make sure that our program is built with the PIO hardware API, so we can call functions like `pio_add_program()` in our C file.

- `pico_add_extra_outputs()`: By default we just get an `.elf` file as the build output of our app. Here we declare we also want extra build formats, like a `.uf2` file which can be dragged and dropped directly onto a Raspberry Pi Pico attached over USB.

Assuming you already have `pico-examples` and the SDK installed on your machine, you can run

```
$ mkdir build
$ cd build
$ cmake ..
$ make hello_pio
```

To build this program.

3.2.2. A Real Example: WS2812 LEDs

The WS2812 LED (sometimes sold as NeoPixel) is an addressable RGB LED. In other words, it's an LED where the red, green and blue components of the light can be individually controlled, and it can be connected in such a way that many WS2812 LEDs can be controlled individually, with only a single control input. Each LED has a pair of power supply terminals, a serial data input, and a serial data output.

When serial data is presented at the LED's input, it takes the first three bytes for itself (red, green, blue) and the remainder is passed along to its serial data output. Often these LEDs are connected in a single long chain, each LED connected to a common power supply, and each LED's data output connected through to the next LED's input. A long burst of serial data to the first in the chain (the one with its data input unconnected) will deposit three bytes of RGB data in each LED, so their colour and brightness can be individually programmed.

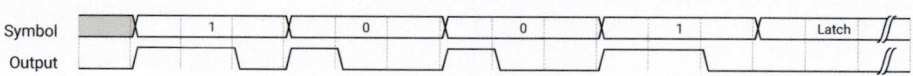

Figure 3. WS2812 line format. Wide positive pulse for 1, narrow positive pulse for 0, very long negative pulse for latch enable

Unfortunately the LEDs receive and retransmit serial data in quite an unusual format. Each bit is transferred as a positive pulse, and the width of the pulse determines whether it is a `1` or a `0` bit. There is a family of WS2812-like LEDs available, which often have slightly different timings, and demand precision. It is possible to bit-bang this protocol, or to write canned bit patterns into some generic serial peripheral like SPI or I2S to get firmer guarantees on the timing, but there is still some software complexity and cost associated with generating the bit patterns.

Ideally we would like to have all of our CPU cycles available to generate colour patterns to put on the lights, or to handle any other responsibilities the processor may have in the *embedded system* the LEDs are connected to.

> **💡 TIP**
>
> Once more, this section is going to discuss a real, complete program, that you can build and run on your Raspberry Pi Pico. Follow the links above the program listings if you'd prefer to build the program yourself and run it, before going through it in detail. This section explores the `pio/ws2812` example in `pico-examples`.

3.2.2.1. PIO Program

Pico Examples: https://github.com/raspberrypi/pico-examples/blob/master/pio/ws2812/ws2812.pio *Lines 7 - 26*

```
 7  .program ws2812
 8  .side_set 1
 9
10  .define public T1 2
11  .define public T2 5
12  .define public T3 3
13
14  .lang_opt python sideset_init = pico.PIO.OUT_HIGH
15  .lang_opt python out_init     = pico.PIO.OUT_HIGH
16  .lang_opt python out_shiftdir = 1
17
18  .wrap_target
19  bitloop:
20      out x, 1       side 0 [T3 - 1] ; Side-set still takes place when instruction stalls
21      jmp !x do_zero side 1 [T1 - 1] ; Branch on the bit we shifted out. Positive pulse
22  do_one:
23      jmp bitloop    side 1 [T2 - 1] ; Continue driving high, for a long pulse
24  do_zero:
25      nop            side 0 [T2 - 1] ; Or drive low, for a short pulse
26  .wrap
```

The previous example was a bit of a whistle-stop tour of the anatomy of a PIO-based application. This time we will dissect the code line-by-line. The first line tells the assembler that we are defining a program named ws2812:

```
.program ws2812
```

We can have multiple programs in one `.pio` file (and you will see this if you click the GitHub link above the main program listing), and each of these will have its own `.program` directive with a different name. The assembler will go through each program in turn, and all the assembled programs will appear in the output file.

Each PIO instruction is 16 bits in size. Generally, 5 of those bits in each instruction are used for the "delay" which is usually 0 to 31 cycles (after the instruction completes and before moving to the next instruction). If you have read the PIO chapter of the **RP2040 Datasheet**, you may have already know that these 5 bits can be used for a different purpose:

```
.side_set 1
```

This directive `.side_set 1` says we're *stealing* one of those delay bits to use for "side-set". The state machine will use this bit to drive the values of some pins, once per instruction, in *addition* to what the instructions are themselves doing. This is very useful for high frequency use cases (e.g. pixel clocks for DPI panels), but also for shrinking program size, to fit into the shared instruction memory.

Note that stealing one bit has left our delay range from 0-15 (4 bits), but that is quite natural because you rarely want to mix side-set with lower frequency stuff. Because we didn't say `.side_set 1 opt`, which would mean the side-set is optional (at the cost of another bit to say *whether* the instruction does a side-set), we have to specify a side-set value for *every* instruction in the program. This is the `side N` you will see on each instruction in the listing.

```
.define public T1 2
.define public T2 5
.define public T3 3
```

`.define` lets you declare constants. The `public` keyword means that the assembler will also write out the value of the define in the output file for use by other software: in the context of the SDK, this is a `#define`. We are going to use `T1`, `T2` and `T3` in calculating the delay cycles on each instruction.

```
.lang_opt python
```

This is used to specify some PIO hardware defaults as used by the MicroPython PIO library. We don't need to worry about them in the context of SDK applications.

```
.wrap_target
```

We'll ignore this for now, and come back to it later, when we meet its friend `.wrap`.

```
bitloop:
```

This is a label. A label tells the assembler that this point in your code is interesting to you, and you want to refer to it later by name. Labels are mainly used with `jmp` instructions.

```
    out x, 1       side 0 [T3 - 1] ; Side-set still takes place when instruction stalls
```

Finally we reach a line with a PIO instruction. There is a lot to see here.

- This is an `out` instruction. `out` takes some bits from the *output shift register* (OSR), and writes them somewhere else. In this case, the OSR will contain pixel data destined for our LEDs.
- `[T3 - 1]` is the number of delay cycles (T3 minus 1). `T3` is a constant we defined earlier.
- `x` (one of two scratch registers; the other imaginatively called `y`) is the destination of the write data. State machines use their scratch registers to hold and compare temporary data.
- `side 0`: Drive low (`0`) the pin configured for side-set.
- Everything after the `;` character is a *comment*. Comments are ignored by the assembler: they are just notes for humans to read.

> **Output Shift Register**
>
> The OSR is a staging area for data entering the state machine through the TX FIFO. Data is pulled from the TX FIFO into the OSR one 32-bit chunk at a time. When an `out` instruction is executed, the OSR can break this data into smaller pieces by *shifting* to the left or right, and sending the bits that drop off the end to one of a handful of different destinations, such as the pins.
>
> The amount of data to be shifted is encoded by the `out` instruction, and the *direction* of the shift (left or right) is configured ahead of time. For full details and diagrams, see the **RP2040 Datasheet**.

So, the state machine will do the following operations when it executes this instruction:

1. Set 0 on the side-set pin (this happens even if the instruction stalls because no data is available in the OSR)
2. Shift one bit out of the OSR into the x register. The value of the x register will be either 0 or 1.
3. Wait `T3 - 1` cycles after the instruction (I.e. the whole thing takes `T3` cycles since the instruction itself took a cycle). Note that when we say cycle, we mean state machine execution cycles: a state machine can be made to execute at a slower rate than the system clock, by configuring its *clock divider*.

Let's look at the next instruction in the program.

```
jmp !x do_zero side 1 [T1 - 1] ; Branch on the bit we shifted out. Positive pulse
```

1. `side 1` on the side-set pin (this is the leading edge of our pulse)
2. If `x == 0` then go to the instruction labelled `do_zero`, otherwise continue on sequentially to the next instruction
3. We delay `T1 - 1` after the instruction (whether the branch is taken or not)

Let's look at what our output pin has done so far in the program.

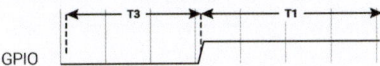

Figure 4. The state machine drives the line low for time T1 as it shifts out one data bit from the OSR, and then high for time T2 whilst branching on the value of the bit.

The pin has been low for time T3, and high for time T1. If the x register is 1 (remember this contains our 1 bit of pixel data) then we will fall through to the instruction labelled `do_one`:

```
do_one:
    jmp  bitloop   side 1 [T2 - 1] ; Continue driving high, for a long pulse
```

On this side of the branch we do the following:

1. `side 1` on the side-set pin (continue the pulse)
2. `jmp` unconditionally back to `bitloop` (the label we defined earlier, at the top of the program); the state machine is done with this data bit, and will get another from its OSR
3. Delay for `T2 - 1` cycles after the instruction

The waveform at our output pin now looks like this:

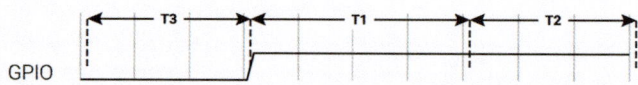

Figure 5. On a one data bit, the line is driven low for time T3, high for time T1, then high for an additional time T2

This accounts for the case where we shifted a `1` data bit into the x register. For a `0` bit, we will have jumped over the last instruction we looked at, to the instruction labelled `do_zero`:

```
do_zero:
    nop             side 0 [T2 - 1] ; Or drive low, for a short pulse
```

1. `side 0` on the side-set pin (the trailing edge of our pulse)
2. `nop` means no operation. We don't have anything else we particularly want to do, so waste a cycle
3. The instruction takes T2 cycles in total

For the `x == 0` case, we get this on our output pin:

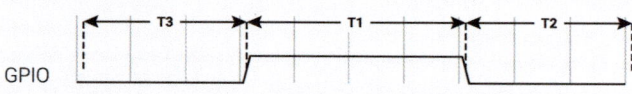

Figure 6. On a zero data bit, the line is driven low for time T3, high for time T1, then low again for time T1

The final line of our program is this:

```
.wrap
```

This matches with the `.wrap_target` directive at the top of the program. Wrapping is a hardware feature of the state machine which behaves like a wormhole: you go in through the `.wrap` statement and appear at the `.wrap_target` *zero* cycles later, unless the `.wrap` is preceded immediately by a `jmp` whose condition is true. This is important for getting precise timing with programs that must run quickly, and often also saves you a slot in the instruction memory.

 TIP

> Often an explicit `.wrap_target`/`.wrap` pair is not necessary, because the default configuration produced by `pioasm` has an implicit wrap from the end of the program back to the beginning, if you didn't specify one.

NOPs

NOP, or no operation, means precisely that: do nothing! You may notice there is no `nop` instruction defined in the instruction set reference: `nop` is really a synonym for `mov y, y` in PIO assembly.

Why did we insert a `nop` in this example when we could have `jmp`-ed? Good question! It's a dramatic device we contrived so we could discuss `nop` and `.wrap`. Writing documentation is hard. In general, though, `nop` is useful when you need to perform a side-set and have nothing else to do, or you need a very slightly longer delay than is available on a single instruction.

It is hopefully becoming clear why our timings T1, T2, T3 are numbered this way, because what the LED string sees really is one of these two cases:

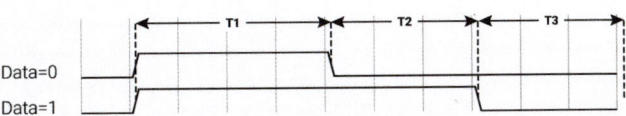

Figure 7. The line is initially low in the idle (latch) state, and the LED is waiting for the first rising edge. It sees our pulse timings in the order T1-T2-T3, until the very last T3, where it sees a much longer negative period once the state machine runs out of data.

This should look familiar if you refer back to Figure 3.

After thoroughly dissecting our program, and hopefully being satisfied that it will repeatedly send one well-formed data bit to a string of WS2812 LEDs, we're left with a question: where is the data coming from? This is more thoroughly explained in the RP2040 Datasheet, but the data that we are shifting out from the OSR came from the state machine's TX FIFO. The TX FIFO is a data buffer between the state machine and the rest of RP2040, filled either via direct poking from the CPU, or by the system DMA, which is much faster.

The `out` instruction shifts data out from the OSR, and zeroes are shifted in from the other end to fill the vacuum. Because the OSR is 32 bits wide, you will start getting zeroes once you have shifted out a total of 32 bits. There is a `pull` instruction which explicitly takes data from the TX FIFO and put it in the OSR (stalling the state machine if the FIFO is empty).

However, in the majority of cases it is simpler to configure *autopull*, a mode where the state machine automatically refills the OSR from the TX FIFO (an automatic `pull`) when a configured number of bits have been shifted out. Autopull happens in the background, in parallel with whatever else the state machine may be up to (in other words it has a cost of zero cycles). We'll see how this is configured in the next section.

3.2.2.2. State Machine Configuration

When we run `pioasm` on the `.pio` file we have been looking at, and ask it to spit out SDK code (which is the default), it will create some static variables describing the program, *and* a method ws2812_default_program_config which configures a PIO state machine based on user parameters, and the directives in the actual PIO program (namely the `.side_set` and `.wrap` in this case).

Of course how you configure the PIO SM when using the program is very much related to the program you have written. Rather than try to store a data representation off all that information, and parse it at runtime, for the use cases where you'd like to encapsulate setup or other API functions with your PIO program, you can embed code within the `.pio` file.

Pico Examples: https://github.com/raspberrypi/pico-examples/blob/master/pio/ws2812/ws2812.pio Lines 31 - 47

```
31  static inline void ws2812_program_init(PIO pio, uint sm, uint offset, uint pin, float freq,
        bool rgbw) {
32
33      pio_gpio_init(pio, pin);
34      pio_sm_set_consecutive_pindirs(pio, sm, pin, 1, true);
35
36      pio_sm_config c = ws2812_program_get_default_config(offset);
37      sm_config_set_sideset_pins(&c, pin);
38      sm_config_set_out_shift(&c, false, true, rgbw ? 32 : 24);
39      sm_config_set_fifo_join(&c, PIO_FIFO_JOIN_TX);
40
41      int cycles_per_bit = ws2812_T1 + ws2812_T2 + ws2812_T3;
42      float div = clock_get_hz(clk_sys) / (freq * cycles_per_bit);
43      sm_config_set_clkdiv(&c, div);
44
45      pio_sm_init(pio, sm, offset, &c);
46      pio_sm_set_enabled(pio, sm, true);
47  }
```

In this case we are passing through code for the SDK, as requested by this line you will see if you click the link on the above listing to see the context:

```
% c-sdk {
```

We have here a function `ws2812_program_init` which is provided to help the user to instantiate an instance of the LED driver program, based on a handful of parameters:

pio

Which of RP2040's two PIO instances we are dealing with

sm

Which state machine on that PIO we want to configure to run the WS2812 program

offset

Where the PIO program was loaded in PIO's 5-bit program address space

pin

which GPIO pin our WS2812 LED chain is connected to

freq

The frequency (or rather baud rate) we want to output data at.

rgbw

True if we are using 4-colour LEDs (red, green, blue, white) rather than the usual 3.

Such that:

- `pio_gpio_init(pio, pin);` Configure a GPIO for use by PIO. (Set the GPIO function select.)

- `pio_set_consecutive_pindirs(pio, sm, pin, 1, true);` Sets the PIO pin direction of 1 pin starting at pin number `pin` to out

- `pio_sm_config c = ws2812_program_default_config(offset);` Get the default configuration using the generated function for this program (this includes things like the `.wrap` and `.side_set` configurations from the program). We'll modify this configuration before loading it into the state machine.

- `sm_config_sideset_pins(&c, pin);` Sets the side-set to write to pins starting at pin `pin` (we say *starting* at because if you had `.side_set` 3, then it would be outputting values on numbers `pin`, `pin+1`, `pin+2`)

- `sm_config_out_shift(&c, false, true, rgbw ? 32 : 24);` False for shift_to_right (i.e. we want to shift out MSB first). True for autopull. 32 or 24 for the number of bits for the autopull threshold, i.e. the point at which the state machine triggers a refill of the OSR, depending on whether the LEDs are RGB or RGBW.

- `int cycles_per_bit = ws2812_T1 + ws2812_T2 + ws2812_T3;` This is the total number of execution cycles to output a single bit. Here we see the benefit of `.define public`; we can use the T1 - T3 values in our code.

- `float div = clock_get_hz(clk_sys) / (freq * cycles_per_bit); sm_config_clkdiv(&c, div);` Slow the state machine's execution down, based on the system clock speed and the number of execution cycles required per WS2812 data bit, so that we achieve the correct bit rate.

- `pio_sm_init(pio, sm, offset, &c);` Load our configuration into the state machine, and go to the start address (`offset`)

- `pio_sm_enable(pio, sm, true);` And make it go now!

At this point the program will be stuck on the first `out` waiting for data. This is because we have autopull enabled, the OSR is initially empty, and there is no data to be pulled. The state machine refuses to continue until the first piece of data arrives in the FIFO.

As an aside, this last point sheds some light on the slightly cryptic comment at the start of the PIO program:

```
        out x, 1      side 0 [T3 - 1] ; Side-set still takes place when instruction stalls
```

This comment is giving us an important piece of context. We stall on this instruction initially, before the first data is added, and also every time we finish sending the last piece of data at the end of a long serial burst. When a state machine stalls, it does not continue to the next instruction, rather it will reattempt the current instruction on the next divided clock cycle. However, side-set still takes place. This works in our favour here, because we consequently always return the line to the idle (low) state when we stall.

3.2.2.3. C Program

The companion to the `.pio` file we've looked at is a `.c` file which drives some interesting colour patterns out onto a string of LEDs. We'll just look at the parts that are directly relevant to PIO.

Pico Examples: https://github.com/raspberrypi/pico-examples/blob/master/pio/ws2812/ws2812.c *Lines 25 - 27*

```
25  static inline void put_pixel(uint32_t pixel_grb) {
26      pio_sm_put_blocking(pio0, 0, pixel_grb << 8u);
27  }
```

Pico Examples: https://github.com/raspberrypi/pico-examples/blob/master/pio/ws2812/ws2812.c *Lines 29 - 34*

```
29  static inline uint32_t urgb_u32(uint8_t r, uint8_t g, uint8_t b) {
30      return
31              ((uint32_t) (r) << 8) |
32              ((uint32_t) (g) << 16) |
33              (uint32_t) (b);
34  }
```

Here we are writing 32-bit values into the FIFO, one at a time, directly from the CPU. `pio_sm_put_blocking` is a helper method that waits until there is room in the FIFO before pushing your data.

You'll notice the `<< 8` in `put_pixel()`: remember we are shifting out starting with the MSB, so we want the 24-bit colour values at the top. This works fine for WGBR too, just that the W is always 0.

This program has a handful of colour patterns, which call our `put_pixel` helper above to output a sequence of pixel values:

Pico Examples: https://github.com/raspberrypi/pico-examples/blob/master/pio/ws2812/ws2812.c *Lines 50 - 55*

```
50  void pattern_random(uint len, uint t) {
51      if (t % 8)
52          return;
53      for (int i = 0; i < len; ++i)
54          put_pixel(rand());
55  }
```

The main function loads the program onto a PIO, configures a state machine for 800 kbaud WS2812 transmission, and then starts cycling through the colour patterns randomly.

Pico Examples: https://github.com/raspberrypi/pico-examples/blob/master/pio/ws2812/ws2812.c *Lines 84 - 108*

```
84  int main() {
85      //set_sys_clock_48();
86      stdio_init_all();
87      printf("WS2812 Smoke Test, using pin %d", WS2812_PIN);
88
89      // todo get free sm
90      PIO pio = pio0;
```

```
 91        int sm = 0;
 92        uint offset = pio_add_program(pio, &ws2812_program);
 93
 94        ws2812_program_init(pio, sm, offset, WS2812_PIN, 800000, IS_RGBW);
 95
 96        int t = 0;
 97        while (1) {
 98            int pat = rand() % count_of(pattern_table);
 99            int dir = (rand() >> 30) & 1 ? 1 : -1;
100            puts(pattern_table[pat].name);
101            puts(dir == 1 ? "(forward)" : "(backward)");
102            for (int i = 0; i < 1000; ++i) {
103                pattern_table[pat].pat(NUM_PIXELS, t);
104                sleep_ms(10);
105                t += dir;
106            }
107        }
108 }
```

3.2.3. PIO and DMA (A Logic Analyser)

So far we have looked at writing data to PIO directly from the processor. This often leads to the processor spinning its wheels waiting for room in a FIFO to make a data transfer, which is not a good investment of its time. It also limits the total data throughput you can achieve.

RP2040 is equipped with a powerful *direct memory access* unit (DMA), which can transfer data for you in the background. Suitably programmed, the DMA can make quite long sequences of transfers without supervision. Up to one word per system clock can be transferred to or from a PIO state machine, which is, to be quite technically precise, more bandwidth than you can shake a stick at. The bandwidth is shared across all state machines, but you can use the full amount on *one* state machine.

Let's take a look at the `logic_analyser` example, which uses PIO to sample some of RP2040's own pins, and capture a logic trace of what is going on there, at full system speed.

Pico Examples: *https://github.com/raspberrypi/pico-examples/blob/master/pio/logic_analyser/logic_analyser.c Lines 40 - 63*

```
40 void logic_analyser_init(PIO pio, uint sm, uint pin_base, uint pin_count, float div) {
41     // Load a program to capture n pins. This is just a single `in pins, n`
42     // instruction with a wrap.
43     uint16_t capture_prog_instr = pio_encode_in(pio_pins, pin_count);
44     struct pio_program capture_prog = {
45             .instructions = &capture_prog_instr,
46             .length = 1,
47             .origin = -1
48     };
49     uint offset = pio_add_program(pio, &capture_prog);
50
51     // Configure state machine to loop over this `in` instruction forever,
52     // with autopush enabled.
53     pio_sm_config c = pio_get_default_sm_config();
54     sm_config_set_in_pins(&c, pin_base);
55     sm_config_set_wrap(&c, offset, offset);
56     sm_config_set_clkdiv(&c, div);
57     // Note that we may push at a < 32 bit threshold if pin_count does not
58     // divide 32. We are using shift-to-right, so the sample data ends up
59     // left-justified in the FIFO in this case, with some zeroes at the LSBs.
60     sm_config_set_in_shift(&c, true, true, bits_packed_per_word(pin_count));
61     sm_config_set_fifo_join(&c, PIO_FIFO_JOIN_RX);
62     pio_sm_init(pio, sm, offset, &c);
63 }
```

Our program consists only of a single `in pins, <pin_count>` instruction, with program wrapping and autopull enabled. Because the amount of data to be shifted is only known at runtime, and because the program is so short, we are generating the program dynamically here (using the `pio_encode_` functions) instead of pushing it through `pioasm`. The program is wrapped in a data structure stating how big the program is, and where it must be loaded — in this case `origin = -1` meaning "don't care".

> **Input Shift Register**
>
> The *input shift register* (ISR) is the mirror image of the OSR. Generally data flows through a state machine in one of two directions: System → TX FIFO → OSR → Pins, or Pins → ISR → RX FIFO → System. An `in` instruction shifts data into the ISR.
>
> If you don't need the ISR's shifting ability — for example, if your program is output-only — you can use the ISR as a third scratch register. It's 32 bits in size, the same as X, Y and the OSR. The full details are in the [RP2040 Datasheet](#).

We load the program into the chosen PIO, and then configure the input pin mapping on the chosen state machine so that its `in pins` instruction will see the pins we care about. For an `in` instruction we only need to worry about configuring the base pin, i.e. the pin which is the least significant bit of the `in` instruction's sample. The number of pins to be sampled is determined by the bit count parameter of the `in pins` instruction — it will sample *n* pins starting at the base we specified, and shift them into the ISR.

> **Pin Groups (Mapping)**
>
> We mentioned earlier that there are four pin groups to configure, to connect a state machine's internal data buses to the GPIOs it manipulates. A state machine accesses all pins within a group at once, and pin groups can overlap. So far we have seen the *out*, *side-set* and *in* pin groups. The fourth is *set*.
>
> The out group is the pins affected by shifting out data from the OSR, using `out pins` or `out pindirs`, up to 32 bits at a time. The set group is used with `set pins` and `set pindirs` instructions, up to 5 bits at a time, with data that is encoded directly in the instruction. It's useful for toggling control signals. The side-set group is similar to the set group, but runs simultaneously with another instruction. Note: `mov pin` uses the in or out group, depending on direction.

Configuring the clock divider optionally slows down the state machine's execution: a clock divisor of *n* means 1 instruction will be executed per *n* system clock cycles. The default system clock frequency for SDK is 125MHz.

`sm_config_set_in_shift` sets the shift direction to rightward, enables autopush, and sets the autopush threshold to 32. The state machine keeps an eye on the total amount of data shifted into the ISR, and on the `in` which reaches or breaches a total shift count of 32 (or whatever number you have configured), the ISR contents, along with the new data from the `in`, goes straight to the RX FIFO. The ISR is cleared to zero in the same operation.

`sm_config_set_fifo_join` is used to manipulate the FIFOs so that the DMA can get more throughput. If we want to sample every pin on every clock cycle, that's a lot of bandwidth! We've finished describing how the state machine should be configured, so we use `pio_sm_init` to load the configuration into the state machine, and get the state machine into a clean initial state.

FIFO Joining

Each state machine is equipped with a FIFO going in each direction: the TX FIFO buffers data on its way out of the system, and the RX FIFO does the same for data coming in. Each FIFO has four data slots, each holding 32 bits of data. Generally you want FIFOs to be as deep as possible, so there is more slack time between the timing-critical operation of a peripheral, and data transfers from system agents which may be quite busy or have high access latency. However this comes with significant hardware cost.

If you are only using one of the two FIFOs — TX or RX — a state machine can pool its resources to provide a single FIFO with double the depth. The RP2040 Datasheet goes into much more detail, including how this mechanism actually works under the hood.

Our state machine is ready to sample some pins. Let's take a look at how we hook up the DMA to our state machine, and tell the state machine to start sampling once it sees some trigger condition.

Pico Examples: https://github.com/raspberrypi/pico-examples/blob/master/pio/logic_analyser/logic_analyser.c Lines 65 - 87

```c
void logic_analyser_arm(PIO pio, uint sm, uint dma_chan, uint32_t *capture_buf, size_t capture_size_words,
                       uint trigger_pin, bool trigger_level) {
    pio_sm_set_enabled(pio, sm, false);
    // Need to clear _input shift counter_, as well as FIFO, because there may be
    // partial ISR contents left over from a previous run. sm_restart does this.
    pio_sm_clear_fifos(pio, sm);
    pio_sm_restart(pio, sm);

    dma_channel_config c = dma_channel_get_default_config(dma_chan);
    channel_config_set_read_increment(&c, false);
    channel_config_set_write_increment(&c, true);
    channel_config_set_dreq(&c, pio_get_dreq(pio, sm, false));

    dma_channel_configure(dma_chan, &c,
        capture_buf,        // Destination pointer
        &pio->rxf[sm],      // Source pointer
        capture_size_words, // Number of transfers
        true                // Start immediately
    );

    pio_sm_exec(pio, sm, pio_encode_wait_gpio(trigger_level, trigger_pin));
    pio_sm_set_enabled(pio, sm, true);
}
```

We want the DMA to read from the RX FIFO on our PIO state machine, so every DMA read is from the same address. The *write* address, on the other hand, should increment after every DMA transfer so that the DMA gradually fills up our capture buffer as data comes in. We need to specify a *data request* signal (DREQ) so that the DMA transfers data at the proper rate.

Data request signals

The DMA can transfer data incredibly fast, and almost invariably this will be much faster than your PIO program actually needs. The DMA paces itself based on a data request handshake with the state machine, so there's no worry about it overflowing or underflowing a FIFO, as long as you have selected the correct DREQ signal. The state machine coordinates with the DMA to tell it when it has room available in its TX FIFO, or data available in its RX FIFO.

We need to provide the DMA channel with an initial read address, an initial write address, and the total number of reads/writes to be performed (*not* the total number of bytes). We start the DMA channel immediately — from this point on, the DMA is poised, waiting for the state machine to produce data. As soon as data appears in the RX FIFO, the DMA will pounce and whisk the data away to our capture buffer in system memory.

As things stand right now, the state machine will immediately go into a 1-cycle loop of `in` instructions once enabled. Since the system memory available for capture is quite limited, it would be better for the state machine to wait for some trigger before it starts sampling. Specifically, we are using a `wait pin` instruction to stall the state machine until a certain pin goes high or low, and again we are using one of the `pio_encode_` functions to encode this instruction on-the-fly.

`pio_sm_exec` tells the state machine to immediately execute some instruction you give it. This instruction never gets written to the instruction memory, and if the instruction stalls (as it will in this case — a `wait` instruction's job is to stall) then the state machine will latch the instruction until it completes. With the state machine stalled on the `wait` instruction, we can enable it without being immediately flooded by data.

At this point everything is armed and waiting for the trigger signal from the chosen GPIO. This will lead to the following sequence of events:

1. The `wait` instruction will clear
2. On the very next cycle, state machine will start to execute `in` instructions from the program memory
3. As soon as data appears in the RX FIFO, the DMA will start to transfer it.
4. Once the requested amount of data has been transferred by the DMA, it'll automatically stop

> **State Machine EXEC Functionality**
>
> So far our state machines have executed instructions from the instruction memory, but there are other options. One is the `SMx_INSTR` register (used by `pio_sm_exec()`): the state machine will immediately execute whatever you write here, momentarily interrupting the current program it's running if necessary. This is useful for poking around inside the state machine from the system side, for initial setup.
>
> The other two options, which use the same underlying hardware, are `out exec` (shift out an instruction from the data being streamed through the OSR, and execute it) and `mov exec` (execute an instruction stashed in e.g. a scratch register). Besides making people's eyes bulge, these are really useful if you want the state machine to perform some data-defined operation at a certain point in an output stream.

The example code provides this cute function for displaying the captured logic trace as ASCII art in a terminal:

Pico Examples: https://github.com/raspberrypi/pico-examples/blob/master/pio/logic_analyser/logic_analyser.c *Lines 89 - 108*

```
 89 void print_capture_buf(const uint32_t *buf, uint pin_base, uint pin_count, uint32_t
    n_samples) {
 90     // Display the capture buffer in text form, like this:
 91     // 00: __--__--__--__--__--__--
 92     // 01: ____----____----____----
 93     printf("Capture:\n");
 94     // Each FIFO record may be only partially filled with bits, depending on
 95     // whether pin_count is a factor of 32.
 96     uint record_size_bits = bits_packed_per_word(pin_count);
 97     for (int pin = 0; pin < pin_count; ++pin) {
 98         printf("%02d: ", pin + pin_base);
 99         for (int sample = 0; sample < n_samples; ++sample) {
100             uint bit_index = pin + sample * pin_count;
101             uint word_index = bit_index / record_size_bits;
102             // Data is left-justified in each FIFO entry, hence the (32 - record_size_bits) offset
103             uint word_mask = 1u << (bit_index % record_size_bits + 32 - record_size_bits);
104             printf(buf[word_index] & word_mask ? "-" : "_");
105         }
106         printf("\n");
107     }
108 }
```

We have everything we need now for RP2040 to capture a logic trace of its own pins, whilst running some other program. Here we're setting up a PWM slice to output at around 15MHz on two GPIOs, and attaching our brand spanking new logic analyser to those same two GPIOs.

Pico Examples: https://github.com/raspberrypi/pico-examples/blob/master/pio/logic_analyser/logic_analyser.c Lines 110 - 159

```c
110 int main() {
111     stdio_init_all();
112     printf("PIO logic analyser example\n");
113
114     // We're going to capture into a u32 buffer, for best DMA efficiency. Need
115     // to be careful of rounding in case the number of pins being sampled
116     // isn't a power of 2.
117     uint total_sample_bits = CAPTURE_N_SAMPLES * CAPTURE_PIN_COUNT;
118     total_sample_bits += bits_packed_per_word(CAPTURE_PIN_COUNT) - 1;
119     uint buf_size_words = total_sample_bits / bits_packed_per_word(CAPTURE_PIN_COUNT);
120     uint32_t *capture_buf = malloc(buf_size_words * sizeof(uint32_t));
121     hard_assert(capture_buf);
122
123     // Grant high bus priority to the DMA, so it can shove the processors out
124     // of the way. This should only be needed if you are pushing things up to
125     // >16bits/clk here, i.e. if you need to saturate the bus completely.
126     bus_ctrl_hw->priority = BUSCTRL_BUS_PRIORITY_DMA_W_BITS | BUSCTRL_BUS_PRIORITY_DMA_R_BITS;
127
128     PIO pio = pio0;
129     uint sm = 0;
130     uint dma_chan = 0;
131
132     logic_analyser_init(pio, sm, CAPTURE_PIN_BASE, CAPTURE_PIN_COUNT, 1.f);
133
134     printf("Arming trigger\n");
135     logic_analyser_arm(pio, sm, dma_chan, capture_buf, buf_size_words, CAPTURE_PIN_BASE, true);
136
137     printf("Starting PWM example\n");
138     // PWM example: ----------------------------------------------------------
139     gpio_set_function(CAPTURE_PIN_BASE, GPIO_FUNC_PWM);
140     gpio_set_function(CAPTURE_PIN_BASE + 1, GPIO_FUNC_PWM);
141     // Topmost value of 3: count from 0 to 3 and then wrap, so period is 4 cycles
142     pwm_hw->slice[0].top = 3;
143     // Divide frequency by two to slow things down a little
144     pwm_hw->slice[0].div = 4 << PWM_CH0_DIV_INT_LSB;
145     // Set channel A to be high for 1 cycle each period (duty cycle 1/4) and
146     // channel B for 3 cycles (duty cycle 3/4)
147     pwm_hw->slice[0].cc =
148             (1 << PWM_CH0_CC_A_LSB) |
149             (3 << PWM_CH0_CC_B_LSB);
150     // Enable this PWM slice
151     pwm_hw->slice[0].csr = PWM_CH0_CSR_EN_BITS;
152     // -----------------------------------------------------------------------
153
154     // The logic analyser should have started capturing as soon as it saw the
155     // first transition. Wait until the last sample comes in from the DMA.
156     dma_channel_wait_for_finish_blocking(dma_chan);
157
158     print_capture_buf(capture_buf, CAPTURE_PIN_BASE, CAPTURE_PIN_COUNT, CAPTURE_N_SAMPLES);
159 }
```

The output of the program looks like this:

```
Starting PWM example
Capture:
16: ----_____----_____----_____----_____----_____
17: ------------____------------____------------____------------____-----------
```

3.2.4. Further examples

Hopefully what you have seen so far has given some idea of how PIO applications can be built with the SDK. The RP2040 Datasheet contains *many* more documented examples, which highlight particular hardware features of PIO, or show how particular hardware interfaces can be implemented.

You can also browse the `pio/` directory in the Pico Examples repository.

3.3. Using PIOASM, the PIO Assembler

Up until now, we have glossed over the details of how the assembly program in our `.pio` file is translated into a binary program, ready to be loaded into our PIO state machine. Programs that handle this task — translating assembly code into binary — are generally referred to as *assemblers*, and PIO is no exception in this regard. The SDK includes an assembler for PIO, called `pioasm`. The SDK handles the details of building this tool for you behind the scenes, and then using it to build your PIO programs, for you to `#include` from your C or C++ program. `pioasm` can also be used directly, and has a few features not used by the C++ SDK, such as generating programs suitable for use with the MicroPython PIO library.

If you have built the `pico-examples` repository at any point, you will likely already have a `pioasm` binary in your build directory, located under `build/tools/pioasm/pioasm`, which was bootstrapped for you before building any applications that depend on it. If we want a standalone copy of `pioasm`, perhaps just to explore the available command-line options, we can obtain it as follows (assuming the SDK is extracted at `$PICO_SDK_PATH`):

```
$ mkdir pioasm_build
$ cd pioasm_build
$ cmake $PICO_SDK_PATH/tools/pioasm
$ make
```

And then invoke as:

```
$ ./pioasm
```

3.3.1. Usage

A description of the command line arguments can be obtained by running:

```
$ pioasm -?
```

giving:

```
usage: pioasm <options> <input> (<output>)

Assemble file of PIO program(s) for use in applications.
<input>                the input filename
<output>               the output filename (or filename prefix if the output
                       format produces multiple outputs).
                    if not specified, a single output will be written to stdout

options:
-o <output_format>   select output_format (default 'c-sdk'); available options are:
                       c-sdk
                           C header suitable for use with the Raspberry Pi Pico SDK
                       python
                           Python file suitable for use with MicroPython
                       hex
                           Raw hex output (only valid for single program inputs)
-p <output_param>    add a parameter to be passed to the outputter
-?, --help           print this help and exit
```

> **NOTE**
>
> Within the SDK you do not need to invoke pioasm directly, as the `CMake` function `pico_generate_pio_header(TARGET PIO_FILE)` takes care of invoking pioasm and adding the generated header to the include path of the target TARGET for you.

3.3.2. Directives

The following directives control the assembly of PIO programs:

Table 5. pioasm directives

Directive	Description
.define (PUBLIC) <symbol> <value>	Define an integer symbol named *<symbol>* with the value *<value>* (see Section 3.3.3). If this *.define* appears before the first program in the input file, then the define is global to all programs, otherwise it is local to the program in which it occurs. If *PUBLIC* is specified the symbol will be emitted into the assembled output for use by user code. For the SDK this takes the form of: `#define <program_name>_<symbol> value` for program symbols or `#define <symbol> value` for global symbols
.program <name>	Start a new program with the name *<name>*. Note that that name is used in code so should be alphanumeric/underscore not starting with a digit. The program lasts until another *.program* directive or the end of the source file. PIO instructions are only allowed within a program
.origin <offset>	Optional directive to specify the PIO instruction memory offset at which the program *must* load. Most commonly this is used for programs that must load at offset 0, because they use data based JMPs with the (absolute) jmp target being stored in only a few bits. This directive is invalid outside of a program
.side_set <count> (opt) (pindirs)	If this directive is present, *<count>* indicates the number of side-set bits to be used. Additionally *opt* may be specified to indicate that a `side <value>` is optional for instructions (note this requires stealing an extra bit — in addition to the *<count>* bits — from those available for the instruction delay). Finally, *pindirs* may be specified to indicate that the side set values should be applied to the PINDIRs and not the PINs. This directive is only valid within a program before the first instruction

.wrap_target	Place prior to an instruction, this directive specifies the instruction where execution continues due to program wrapping. This directive is invalid outside of a program, may only be used once within a program, and if not specified defaults to the start of the program
.wrap	Placed after an instruction, this directive specifies the instruction after which, in normal control flow (i.e. `jmp` with false condition, or no `jmp`), the program wraps (to .wrap_target instruction). This directive is invalid outside of a program, may only be used once within a program, and if not specified defaults to after the last program instruction.
.lang_opt <lang> <name> <option>	Specifies an option for the program related to a particular language generator. (See Section 3.3.10). This directive is invalid outside of a program
.word <value>	Stores a raw 16-bit value as an instruction in the program. This directive is invalid outside of a program.

3.3.3. Values

The following types of values can be used to define integer numbers or branch targets

Table 6. Values in pioasm, i.e. <value>

integer	An integer value e.g. 3 or -7
hex	A hexadecimal value e.g. `0xf`
binary	A binary value e.g. `0b1001`
symbol	A value defined by a `.define` (see [pioasm_define])
<label>	The instruction offset of the label within the program. This makes most sense when used with a JMP instruction (see Section 3.4.2)
(<expression>)	An expression to be evaluated; see expressions. Note that the parentheses are necessary.

3.3.4. Expressions

Expressions may be freely used within pioasm values.

Table 7. Expressions in pioasm i.e. <expression>

<expression> + <expression>	The sum of two expressions
<expression> - <expression>	The difference of two expressions
<expression> * <expression>	The multiplication of two expressions
<expression> / <expression>	The integer division of two expressions
- <expression>	The negation of another expression
:: <expression>	The bit reverse of another expression
<value>	Any value (see Section 3.3.3)

3.3.5. Comments

Line comments are supported with `//` or `;`

C-style block comments are supported via `/*` and `*/`

3.3.6. Labels

Labels are of the form:

`<symbol>:`

or

`PUBLIC <symbol>:`

at the start of a line.

 TIP

> A label is really just an automatic `.define` with a value set to the current program instruction offset. A *PUBLIC* label is exposed to the user code in the same way as a *PUBLIC* `.define`.

3.3.7. Instructions

All pioasm instructions follow a common pattern:

<instruction> (side *<side_set_value>*) ([*<delay_value>*])

where:

<instruction>   Is an assembly instruction detailed in the following sections. (See Section 3.4)

<side_set_value>   Is a value (see Section 3.3.3) to apply to the side_set pins at the start of the instruction. Note that the rules for a side-set value via `side <side_set_value>` are dependent on the `.side_set` (see [pioasm_side_set]) directive for the program. If no `.side_set` is specified then the `side <side_set_value>` is invalid, if an optional number of sideset pins is specified then `side <side_set_value>` may be present, and if a non-optional number of sideset pins is specified, then `side <side_set_value>` is required. The *<side_set_value>* must fit within the number of side-set bits specified in the `.side_set` directive.

<delay_value>   Specifies the number of cycles to delay after the instruction completes. The delay_value is specified as a value (see Section 3.3.3), and in general is between 0 and 31 inclusive (a 5-bit value), however the number of bits is reduced when sideset is enabled via the `.side_set` (see [pioasm_side_set]) directive. If the *<delay_value>* is not present, then the instruction has no delay

 NOTE

> pioasm instruction names, keywords and directives are case insensitive; lower case is used in the *Assembly Syntax* sections below as this is the style used in the SDK.

 NOTE

> Commas appear in some *Assembly Syntax* sections below, but are entirely optional, e.g. `out pins, 3` may be written `out pins 3`, and `jmp x-- label` may be written as `jmp x--, label`. The *Assembly Syntax* sections below uses the first style in each case as this is the style used in the SDK.

3.3.8. Pseudoinstructions

Currently pioasm provides one pseudoinstruction, as a convenience:

nop	Assembles to `mov y, y`. "No operation", has no particular side effect, but a useful vehicle for a side-set operation or an extra delay.

3.3.9. Output pass through

Text in the PIO file may be passed, unmodified, to the output based on the language generator being used.

For example the following (comment and function) would be included in the generated header when the default `c-sdk` language generator is used.

```
% c-sdk {

// an inline function (since this is going in a header file)
static inline int some_c_code() {
    return 0;
}
%}
```

The general format is

```
% target {
pass through contents
%}
```

with `targets` being recognized by a particular language generator (see Section 3.3.10; note that `target` is usually the language generator name e.g. `c-sdk`, but could potentially be `some_language.some_group` if the language generator supports different classes of pass through with different output locations.

This facility allows you to encapsulate both the PIO program and the associated setup required in the same source file. See Section 3.3.10 for a more complete example.

3.3.10. Language generators

The following example shows a multi program source file (with multiple programs) which we will use to highlight c-sdk and python output features

Pico Examples: https://github.com/raspberrypi/pico-examples/blob/master/pio/ws2812/ws2812.pio

```
 1 ;
 2 ; Copyright (c) 2020 Raspberry Pi (Trading) Ltd.
 3 ;
 4 ; SPDX-License-Identifier: BSD-3-Clause
 5 ;
 6
 7 .program ws2812
 8 .side_set 1
 9
10 .define public T1 2
11 .define public T2 5
12 .define public T3 3
13
14 .lang_opt python sideset_init = pico.PIO.OUT_HIGH
15 .lang_opt python out_init     = pico.PIO.OUT_HIGH
16 .lang_opt python out_shiftdir = 1
17
```

```
18 .wrap_target
19 bitloop:
20     out x, 1       side 0 [T3 - 1] ; Side-set still takes place when instruction stalls
21     jmp !x do_zero side 1 [T1 - 1] ; Branch on the bit we shifted out. Positive pulse
22 do_one:
23     jmp bitloop    side 1 [T2 - 1] ; Continue driving high, for a long pulse
24 do_zero:
25     nop            side 0 [T2 - 1] ; Or drive low, for a short pulse
26 .wrap
27
28 % c-sdk {
29 #include "hardware/clocks.h"
30
31 static inline void ws2812_program_init(PIO pio, uint sm, uint offset, uint pin, float freq, bool rgbw) {
32
33     pio_gpio_init(pio, pin);
34     pio_sm_set_consecutive_pindirs(pio, sm, pin, 1, true);
35
36     pio_sm_config c = ws2812_program_get_default_config(offset);
37     sm_config_set_sideset_pins(&c, pin);
38     sm_config_set_out_shift(&c, false, true, rgbw ? 32 : 24);
39     sm_config_set_fifo_join(&c, PIO_FIFO_JOIN_TX);
40
41     int cycles_per_bit = ws2812_T1 + ws2812_T2 + ws2812_T3;
42     float div = clock_get_hz(clk_sys) / (freq * cycles_per_bit);
43     sm_config_set_clkdiv(&c, div);
44
45     pio_sm_init(pio, sm, offset, &c);
46     pio_sm_set_enabled(pio, sm, true);
47 }
48 %}
49
50 .program ws2812_parallel
51
52 .define public T1 2
53 .define public T2 5
54 .define public T3 3
55
56 .wrap_target
57     out x, 32
58     mov pins, !null [T1-1]
59     mov pins, x     [T2-1]
60     mov pins, null  [T3-2]
61 .wrap
62
63 % c-sdk {
64 #include "hardware/clocks.h"
65
66 static inline void ws2812_parallel_program_init(PIO pio, uint sm, uint offset, uint pin_base, uint pin_count, float freq) {
67     for(uint i=pin_base; i<pin_base+pin_count; i++) {
68         pio_gpio_init(pio, i);
69     }
70     pio_sm_set_consecutive_pindirs(pio, sm, pin_base, pin_count, true);
71
72     pio_sm_config c = ws2812_parallel_program_get_default_config(offset);
73     sm_config_set_out_shift(&c, true, true, 32);
74     sm_config_set_out_pins(&c, pin_base, pin_count);
75     sm_config_set_set_pins(&c, pin_base, pin_count);
76     sm_config_set_fifo_join(&c, PIO_FIFO_JOIN_TX);
77
78     int cycles_per_bit = ws2812_parallel_T1 + ws2812_parallel_T2 + ws2812_parallel_T3;
```

```
79      float div = clock_get_hz(clk_sys) / (freq * cycles_per_bit);
80      sm_config_set_clkdiv(&c, div);
81
82      pio_sm_init(pio, sm, offset, &c);
83      pio_sm_set_enabled(pio, sm, true);
84  }
85  %}
```

3.3.10.1. c-sdk

The c-sdk language generator produces a single header file with all the programs in the PIO source file:

The pass through sections (`% c-sdk {`) are embedded in the output, and the `PUBLIC` defines are available via `#define`

> 💡 **TIP**
>
> `pioasm` creates a function for each program (e.g. `ws2812_program_get_default_config()`) returning a `pio_sm_config` based on the `.side_set`, `.wrap` and `.wrap_target` settings of the program, which you can then use as a basis for configuration the PIO state machine.

Pico Examples: *https://github.com/raspberrypi/pico-examples/blob/master/pio/ws2812/generated/ws2812.pio.h*

```
 1  // ------------------------------------------------ //
 2  // This file is autogenerated by pioasm; do not edit! //
 3  // ------------------------------------------------ //
 4
 5  #pragma once
 6
 7  #if !PICO_NO_HARDWARE
 8  #include "hardware/pio.h"
 9  #endif
10
11  // ------ //
12  // ws2812 //
13  // ------ //
14
15  #define ws2812_wrap_target 0
16  #define ws2812_wrap 3
17
18  #define ws2812_T1 2
19  #define ws2812_T2 5
20  #define ws2812_T3 3
21
22  static const uint16_t ws2812_program_instructions[] = {
23              //     .wrap_target
24      0x6221, //  0: out    x, 1            side 0 [2]
25      0x1123, //  1: jmp    !x, 3           side 1 [1]
26      0x1400, //  2: jmp    0               side 1 [4]
27      0xa442, //  3: nop                    side 0 [4]
28              //     .wrap
29  };
30
31  #if !PICO_NO_HARDWARE
32  static const struct pio_program ws2812_program = {
33      .instructions = ws2812_program_instructions,
34      .length = 4,
35      .origin = -1,
36  };
37
```

```c
38  static inline pio_sm_config ws2812_program_get_default_config(uint offset) {
39      pio_sm_config c = pio_get_default_sm_config();
40      sm_config_set_wrap(&c, offset + ws2812_wrap_target, offset + ws2812_wrap);
41      sm_config_set_sideset(&c, 1, false, false);
42      return c;
43  }
44
45  #include "hardware/clocks.h"
46  static inline void ws2812_program_init(PIO pio, uint sm, uint offset, uint pin, float freq,
    bool rgbw) {
47      pio_gpio_init(pio, pin);
48      pio_sm_set_consecutive_pindirs(pio, sm, pin, 1, true);
49      pio_sm_config c = ws2812_program_get_default_config(offset);
50      sm_config_set_sideset_pins(&c, pin);
51      sm_config_set_out_shift(&c, false, true, rgbw ? 32 : 24);
52      sm_config_set_fifo_join(&c, PIO_FIFO_JOIN_TX);
53      int cycles_per_bit = ws2812_T1 + ws2812_T2 + ws2812_T3;
54      float div = clock_get_hz(clk_sys) / (freq * cycles_per_bit);
55      sm_config_set_clkdiv(&c, div);
56      pio_sm_init(pio, sm, offset, &c);
57      pio_sm_set_enabled(pio, sm, true);
58  }
59
60  #endif
61
62  // --------------- //
63  // ws2812_parallel //
64  // --------------- //
65
66  #define ws2812_parallel_wrap_target 0
67  #define ws2812_parallel_wrap 3
68
69  #define ws2812_parallel_T1 2
70  #define ws2812_parallel_T2 5
71  #define ws2812_parallel_T3 3
72
73  static const uint16_t ws2812_parallel_program_instructions[] = {
74              //     .wrap_target
75      0x6020, // 0: out    x, 32
76      0xa10b, // 1: mov    pins, !null            [1]
77      0xa401, // 2: mov    pins, x                [4]
78      0xa103, // 3: mov    pins, null             [1]
79              //     .wrap
80  };
81
82  #if !PICO_NO_HARDWARE
83  static const struct pio_program ws2812_parallel_program = {
84      .instructions = ws2812_parallel_program_instructions,
85      .length = 4,
86      .origin = -1,
87  };
88
89  static inline pio_sm_config ws2812_parallel_program_get_default_config(uint offset) {
90      pio_sm_config c = pio_get_default_sm_config();
91      sm_config_set_wrap(&c, offset + ws2812_parallel_wrap_target, offset +
    ws2812_parallel_wrap);
92      return c;
93  }
94
95  #include "hardware/clocks.h"
96  static inline void ws2812_parallel_program_init(PIO pio, uint sm, uint offset, uint
    pin_base, uint pin_count, float freq) {
97      for(uint i=pin_base; i<pin_base+pin_count; i++) {
```

```
 98              pio_gpio_init(pio, i);
 99          }
100          pio_sm_set_consecutive_pindirs(pio, sm, pin_base, pin_count, true);
101          pio_sm_config c = ws2812_parallel_program_get_default_config(offset);
102          sm_config_set_out_shift(&c, true, true, 32);
103          sm_config_set_out_pins(&c, pin_base, pin_count);
104          sm_config_set_set_pins(&c, pin_base, pin_count);
105          sm_config_set_fifo_join(&c, PIO_FIFO_JOIN_TX);
106          int cycles_per_bit = ws2812_parallel_T1 + ws2812_parallel_T2 + ws2812_parallel_T3;
107          float div = clock_get_hz(clk_sys) / (freq * cycles_per_bit);
108          sm_config_set_clkdiv(&c, div);
109          pio_sm_init(pio, sm, offset, &c);
110          pio_sm_set_enabled(pio, sm, true);
111      }
112
113      #endif
```

3.3.10.2. python

The python language generator produces a single python file with all the programs in the PIO source file:

The pass through sections (`% python {`) would be embedded in the output, and the `PUBLIC` defines are available as python variables.

Also note the use of `.lang_opt python` to pass initializers for the `@pico.asm_pio` decorator

> 💡 **TIP**
>
> The python language output is provided as a utility. MicroPython supports programming with the PIO natively, so you may only want to use pioasm when sharing PIO code between the SDK and MicroPython. No effort is currently made to preserve label names, symbols or comments, as it is assumed you are either using the PIO file as a source or python; not both. The python language output can of course be used to bootstrap your MicroPython PIO development based on an existing PIO file.

Pico Examples: *https://github.com/raspberrypi/pico-examples/blob/master/pio/ws2812/generated/ws2812.py*

```
 1  # ------------------------------------------------ #
 2  # This file is autogenerated by pioasm; do not edit! #
 3  # ------------------------------------------------ #
 4
 5  import rp2
 6  from machine import Pin
 7  # ------ #
 8  # ws2812 #
 9  # ------ #
10
11  ws2812_T1 = 2
12  ws2812_T2 = 5
13  ws2812_T3 = 3
14
15  @rp2.asm_pio(sideset_init=pico.PIO.OUT_HIGH, out_init=pico.PIO.OUT_HIGH, out_shiftdir=1)
16  def ws2812():
17      wrap_target()
18      label("0")
19      out(x, 1)                  .side(0) [2]   # 0
20      jmp(not_x, "3")            .side(1) [1]   # 1
21      jmp("0")                   .side(1) [4]   # 2
22      label("3")
23      nop()                      .side(0) [4]   # 3
```

```
24      wrap()
25
26
27
28 # --------------- #
29 # ws2812_parallel #
30 # --------------- #
31
32 ws2812_parallel_T1 = 2
33 ws2812_parallel_T2 = 5
34 ws2812_parallel_T3 = 3
35
36 @rp2.asm_pio()
37 def ws2812_parallel():
38     wrap_target()
39     out(x, 32)                         # 0
40     mov(pins, invert(null))     [1]    # 1
41     mov(pins, x)                [4]    # 2
42     mov(pins, null)             [1]    # 3
43     wrap()
```

3.3.10.3. hex

The hex generator only supports a single input program, as it just dumps the raw instructions (one per line) as a 4-character hexadecimal number.

Given:

Pico Examples: *https://github.com/raspberrypi/pico-examples/blob/master/pio/squarewave/squarewave.pio*

```
 1 ;
 2 ; Copyright (c) 2020 Raspberry Pi (Trading) Ltd.
 3 ;
 4 ; SPDX-License-Identifier: BSD-3-Clause
 5 ;
 6
 7 .program squarewave
 8     set pindirs, 1   ; Set pin to output
 9 again:
10     set pins, 1 [1]  ; Drive pin high and then delay for one cycle
11     set pins, 0      ; Drive pin low
12     jmp again        ; Set PC to label `again`
```

The *hex* output produces:

Pico Examples: *https://github.com/raspberrypi/pico-examples/blob/master/pio/squarewave/generated/squarewave.hex*

```
1 e081
2 e101
3 e000
4 0001
```

3.4. PIO Instruction Set Reference

> **ℹ NOTE**
>
> This section refers in places to concepts and pieces of hardware discussed in the RP2040 Datasheet. You are encouraged to read the PIO chapter of the datasheet to get the full context for what these instructions do.

3.4.1. Summary

PIO instructions are 16 bits long, and have the following encoding:

Table 8. PIO instruction encoding

Bit:	15	14	13	12	11	10	9	8	7	6	5	4	3	2	1	0
JMP	0	0	0	Delay/side-set					Condition			Address			
WAIT	0	0	1	Delay/side-set					Pol	Source		Index			
IN	0	1	0	Delay/side-set					Source			Bit count			
OUT	0	1	1	Delay/side-set					Destination			Bit count			
PUSH	1	0	0	Delay/side-set					0	IfF	Blk	0	0	0	0	0
PULL	1	0	0	Delay/side-set					1	IfE	Blk	0	0	0	0	0
MOV	1	0	1	Delay/side-set					Destination			Op		Source		
IRQ	1	1	0	Delay/side-set					0	Clr	Wait	Index			
SET	1	1	1	Delay/side-set					Destination			Data				

All PIO instructions execute in one clock cycle.

The `Delay/side-set` field is present in all instructions. Its exact use is configured for each state machine by `PINCTRL_SIDESET_COUNT`:

- Up to 5 MSBs encode a side-set operation, which optionally asserts a constant value onto some GPIOs, concurrently with main instruction execution logic
- Remaining LSBs (up to 5) encode the number of idle cycles inserted between this instruction and the next

3.4.2. JMP

3.4.2.1. Encoding

Bit:	15	14	13	12	11	10	9	8	7	6	5	4	3	2	1	0
JMP	0	0	0	Delay/side-set					Condition			Address			

3.4.2.2. Operation

Set program counter to `Address` if `Condition` is true, otherwise no operation.

Delay cycles on a `JMP` always take effect, whether `Condition` is true or false, and they take place *after* `Condition` is evaluated and the program counter is updated.

- Condition:
    - 000: *(no condition)*: Always
    - 001: `!X`: scratch X zero

- 010: `X--`: scratch X non-zero, post-decrement
- 011: `!Y`: scratch Y zero
- 100: `Y--`: scratch Y non-zero, post-decrement
- 101: `X!=Y`: scratch X not equal scratch Y
- 110: `PIN`: branch on input pin
- 111: `!OSRE`: output shift register not empty

- Address: Instruction address to jump to. In the instruction encoding this is an absolute address within the PIO instruction memory.

`JMP PIN` branches on the GPIO selected by `EXECCTRL_JMP_PIN`, a configuration field which selects one out of the maximum of 32 GPIO inputs visible to a state machine, independently of the state machine's other input mapping. The branch is taken if the GPIO is high.

`!OSRE` compares the bits shifted out since the last `PULL` with the shift count threshold configured by `SHIFTCTRL_PULL_THRESH`. This is the same threshold used by autopull.

`JMP X--` and `JMP Y--` always decrement scratch register X or Y, respectively. The decrement is not conditional on the current value of the scratch register. The branch is conditioned on the *initial* value of the register, i.e. before the decrement took place: if the register is initially nonzero, the branch is taken.

3.4.2.3. Assembler Syntax

jmp (<cond>) <target>

where:

<cond>	Is an optional condition listed above (e.g. `!x` for scratch X zero). If a condition code is not specified, the branch is always taken
<target>	Is a program label or value (see Section 3.3.3) representing instruction offset within the program (the first instruction being offset 0). Note that because the PIO JMP instruction uses absolute addresses in the PIO instruction memory, JMPs need to be adjusted based on the program load offset at runtime. This is handled for you when loading a program with the SDK, but care should be taken when encoding JMP instructions for use by `OUT EXEC`

3.4.3. WAIT

3.4.3.1. Encoding

Bit:	15	14	13	12	11	10	9	8	7	6	5	4	3	2	1	0
`WAIT`	0	0	1	Delay/side-set					Pol	Source		Index			

3.4.3.2. Operation

Stall until some condition is met.

Like all stalling instructions, delay cycles begin after the instruction *completes*. That is, if any delay cycles are present, they do not begin counting until *after* the wait condition is met.

- Polarity:

- 1: wait for a 1.
- 0: wait for a 0.

- Source: what to wait on. Values are:
  - 00: `GPIO`: System GPIO input selected by `Index`. This is an absolute GPIO index, and is not affected by the state machine's input IO mapping.
  - 01: `PIN`: Input pin selected by `Index`. This state machine's input IO mapping is applied first, and then `Index` selects which of the mapped bits to wait on. In other words, the pin is selected by adding `Index` to the `PINCTRL_IN_BASE` configuration, modulo 32.
  - 10: `IRQ`: PIO IRQ flag selected by `Index`
  - 11: Reserved
- Index: which pin or bit to check.

`WAIT x IRQ` behaves slightly differently from other `WAIT` sources:

- If `Polarity` is 1, the selected IRQ flag is cleared by the state machine upon the wait condition being met.
- The flag index is decoded in the same way as the `IRQ` index field: if the MSB is set, the state machine ID (0…3) is added to the IRQ index, by way of modulo-4 addition on the two LSBs. For example, state machine 2 with a flag value of '0x11' will wait on flag 3, and a flag value of '0x13' will wait on flag 1. This allows multiple state machines running the same program to synchronise with each other.

> ⚠ **CAUTION**
>
> `WAIT 1 IRQ x` should not be used with IRQ flags presented to the interrupt controller, to avoid a race condition with a system interrupt handler

3.4.3.3. Assembler Syntax

wait <polarity> gpio <gpio_num>

wait <polarity> pin <pin_num>

wait <polarity> irq <irq_num> (rel)

where:

<polarity>	Is a value (see Section 3.3.3) specifying the polarity (either 0 or 1)
<pin_num>	Is a value (see Section 3.3.3) specifying the input pin number (as mapped by the SM input pin mapping)
<gpio_num>	Is a value (see Section 3.3.3) specifying the actual GPIO pin number
<irq_num> (*rel*)	Is a value (see Section 3.3.3) specifying The irq number to wait on (0-7). If *rel* is present, then the actual irq number used is calculating by replacing the low two bits of the irq number (irq_num_{10}) with the low two bits of the sum ($irq\_num_{10} + sm\_num_{10}$) where sm_num_{10} is the state machine number

3.4.4. IN

3.4.4.1. Encoding

Bit:	15	14	13	12	11	10	9	8	7	6	5	4	3	2	1	0
IN	0	1	0	Delay/side-set					Source			Bit count				

3.4.4.2. Operation

Shift `Bit count` bits from `Source` into the Input Shift Register (ISR). Shift direction is configured for each state machine by `SHIFTCTRL_IN_SHIFTDIR`. Additionally, increase the input shift count by `Bit count`, saturating at 32.

- Source:
    - 000: `PINS`
    - 001: `X` (scratch register X)
    - 010: `Y` (scratch register Y)
    - 011: `NULL` (all zeroes)
    - 100: Reserved
    - 101: Reserved
    - 110: `ISR`
    - 111: `OSR`
- Bit count: How many bits to shift into the ISR. 1…32 bits, 32 is encoded as `00000`.

If automatic push is enabled, `IN` will also push the ISR contents to the RX FIFO if the push threshold is reached (`SHIFTCTRL_PUSH_THRESH`). `IN` still executes in one cycle, whether an automatic push takes place or not. The state machine will stall if the RX FIFO is full when an automatic push occurs. An automatic push clears the ISR contents to all-zeroes, and clears the input shift count.

`IN` always uses the least significant `Bit count` bits of the source data. For example, if `PINCTRL_IN_BASE` is set to 5, the instruction `IN PINS, 3` will take the values of pins 5, 6 and 7, and shift these into the ISR. First the ISR is shifted to the left or right to make room for the new input data, then the input data is copied into the gap this leaves. The bit order of the input data is not dependent on the shift direction.

`NULL` can be used for shifting the ISR's contents. For example, UARTs receive the LSB first, so must shift to the right. After 8 `IN PINS, 1` instructions, the input serial data will occupy bits 31…24 of the ISR. An `IN NULL, 24` instruction will shift in 24 zero bits, aligning the input data at ISR bits 7…0. Alternatively, the processor or DMA could perform a byte read from FIFO address + 3, which would take bits 31…24 of the FIFO contents.

3.4.4.3. Assembler Syntax

in <source>, <bit_count>

where:

<source>   Is one of the sources specified above.

<bit_count>   Is a value (see Section 3.3.3) specifying the number of bits to shift (valid range 1-32)

3.4.5. OUT

3.4.5.1. Encoding

Bit:	15	14	13	12	11	10	9	8	7	6	5	4	3	2	1	0
OUT	0	1	1	\multicolumn{5}{c}{Delay/side-set}					Destination			Bit count		

3.4.5.2. Operation

Shift `Bit count` bits out of the Output Shift Register (OSR), and write those bits to `Destination`. Additionally, increase the output shift count by `Bit count`, saturating at 32.

- Destination:
    - 000: `PINS`
    - 001: `X` (scratch register X)
    - 010: `Y` (scratch register Y)
    - 011: `NULL` (discard data)
    - 100: `PINDIRS`
    - 101: `PC`
    - 110: `ISR` (also sets ISR shift counter to `Bit count`)
    - 111: `EXEC` (Execute OSR shift data as instruction)
- Bit count: how many bits to shift out of the OSR. 1…32 bits, 32 is encoded as `00000`.

A 32-bit value is written to `Destination`: the lower `Bit count` bits come from the OSR, and the remainder are zeroes. This value is the least significant `Bit count` bits of the OSR if `SHIFTCTRL_OUT_SHIFTDIR` is to the right, otherwise it is the most significant bits.

`PINS` and `PINDIRS` use the `OUT` pin mapping.

If automatic pull is enabled, the OSR is automatically refilled from the TX FIFO if the pull threshold, `SHIFTCTRL_PULL_THRESH`, is reached. The output shift count is simultaneously cleared to 0. In this case, the `OUT` will stall if the TX FIFO is empty, but otherwise still executes in one cycle.

`OUT EXEC` allows instructions to be included inline in the FIFO datastream. The `OUT` itself executes on one cycle, and the instruction from the OSR is executed on the next cycle. There are no restrictions on the types of instructions which can be executed by this mechanism. Delay cycles on the initial `OUT` are ignored, but the executee may insert delay cycles as normal.

`OUT PC` behaves as an unconditional jump to an address shifted out from the OSR.

3.4.5.3. Assembler Syntax

out <destination>, <bit_count>

where:

<destination>   Is one of the destinations specified above.

<bit_count>   Is a value (see Section 3.3.3) specifying the number of bits to shift (valid range 1-32)

3.4.6. PUSH

3.4.6.1. Encoding

Bit:	15	14	13	12	11	10	9	8	7	6	5	4	3	2	1	0
PUSH	1	0	0		Delay/side-set				0	IfF	Blk	0	0	0	0	0

3.4.6.2. Operation

Push the contents of the ISR into the RX FIFO, as a single 32-bit word. Clear ISR to all-zeroes.

- `IfFull`: If 1, do nothing unless the total input shift count has reached its threshold, `SHIFTCTRL_PUSH_THRESH` (the same as for autopush).
- `Block`: If 1, stall execution if RX FIFO is full.

`PUSH IFFULL` helps to make programs more compact, like autopush. It is useful in cases where the `IN` would stall at an inappropriate time if autopush were enabled, e.g. if the state machine is asserting some external control signal at this point.

The PIO assembler sets the `Block` bit by default. If the `Block` bit is not set, the `PUSH` does not stall on a full RX FIFO, instead continuing immediately to the next instruction. The FIFO state and contents are unchanged when this happens. The ISR is still cleared to all-zeroes, and the `FDEBUG_RXSTALL` flag is set (the same as a blocking `PUSH` or autopush to a full RX FIFO) to indicate data was lost.

3.4.6.3. Assembler Syntax

push (iffull)

push (iffull) block

push (iffull) noblock

where:

iffull	Is equivalent to `IfFull == 1` above. i.e. the default if this is not specified is `IfFull == 0`
block	Is equivalent to `Block == 1` above. This is the default if neither *block* nor *noblock* are specified
noblock	Is equivalent to `Block == 0` above.

3.4.7. PULL

3.4.7.1. Encoding

Bit:	15	14	13	12	11	10	9	8	7	6	5	4	3	2	1	0
PULL	1	0	0		Delay/side-set				1	IfE	Blk	0	0	0	0	0

3.4.7.2. Operation

Load a 32-bit word from the TX FIFO into the OSR.

- `IfEmpty`: If 1, do nothing unless the total output shift count has reached its threshold, `SHIFTCTRL_PULL_THRESH` (the same as for autopull).
- `Block`: If 1, stall if TX FIFO is empty. If 0, pulling from an empty FIFO copies scratch X to OSR.

Some peripherals (UART, SPI...) should halt when no data is available, and pick it up as it comes in; others (I2S) should clock continuously, and it is better to output placeholder or repeated data than to stop clocking. This can be achieved with the `Block` parameter.

A nonblocking `PULL` on an empty FIFO has the same effect as `MOV OSR, X`. The program can either preload scratch register X with a suitable default, or execute a `MOV X, OSR` after each `PULL NOBLOCK`, so that the last valid FIFO word will be recycled until new data is available.

`PULL IFEMPTY` is useful if an `OUT` with autopull would stall in an inappropriate location when the TX FIFO is empty. For example, a UART transmitter should not stall immediately after asserting the start bit. `IfEmpty` permits some of the same program simplifications as autopull, but the stall occurs at a controlled point in the program.

 NOTE

When autopull is enabled, any `PULL` instruction is a no-op when the OSR is full, so that the `PULL` instruction behaves as a barrier. `OUT NULL, 32` can be used to explicitly discard the OSR contents. See the RP2040 Datasheet for more detail on autopull.

3.4.7.3. Assembler Syntax

pull (*ifempty*)

pull (*ifempty*) block

pull (*ifempty*) noblock

where:

ifempty	Is equivalent to `IfEmpty == 1` above. i.e. the default if this is not specified is `IfEmpty == 0`
block	Is equivalent to `Block == 1` above. This is the default if neither *block* nor *noblock* are specified
noblock	Is equivalent to `Block == 0` above.

3.4.8. MOV

3.4.8.1. Encoding

Bit:	15	14	13	12	11	10	9	8	7	6	5	4	3	2	1	0
MOV	1	0	1	\multicolumn{5}{c}{Delay/side-set}					Destination			Op		Source		

3.4.8.2. Operation

Copy data from `Source` to `Destination`.

- Destination:
    - 000: `PINS` (Uses same pin mapping as `OUT`)
    - 001: `X` (Scratch register X)
    - 010: `Y` (Scratch register Y)
    - 011: Reserved
    - 100: `EXEC` (Execute data as instruction)

- 101: `PC`
- 110: `ISR` (Input shift counter is reset to 0 by this operation, i.e. empty)
- 111: `OSR` (Output shift counter is reset to 0 by this operation, i.e. full)

• Operation:
  - 00: None
  - 01: Invert (bitwise complement)
  - 10: Bit-reverse
  - 11: Reserved

• Source:
  - 000: `PINS` (Uses same pin mapping as `IN`)
  - 001: `X`
  - 010: `Y`
  - 011: `NULL`
  - 100: Reserved
  - 101: `STATUS`
  - 110: `ISR`
  - 111: `OSR`

`MOV PC` causes an unconditional jump. `MOV EXEC` has the same behaviour as `OUT EXEC` (Section 3.4.5), and allows register contents to be executed as an instruction. The `MOV` itself executes in 1 cycle, and the instruction in `Source` on the next cycle. Delay cycles on `MOV EXEC` are ignored, but the executee may insert delay cycles as normal.

The `STATUS` source has a value of all-ones or all-zeroes, depending on some state machine status such as FIFO full/empty, configured by `EXECCTRL_STATUS_SEL`.

`MOV` can manipulate the transferred data in limited ways, specified by the `Operation` argument. Invert sets each bit in `Destination` to the logical NOT of the corresponding bit in `Source`, i.e. 1 bits become 0 bits, and vice versa. Bit reverse sets each bit n in `Destination` to bit 31 - n in `Source`, assuming the bits are numbered 0 to 31.

`MOV dst, PINS` reads pins using the `IN` pin mapping, and writes the full 32-bit value to the destination without masking. The LSB of the read value is the pin indicated by `PINCTRL_IN_BASE`, and each successive bit comes from a higher-numbered pin, wrapping after 31.

3.4.8.3. Assembler Syntax

mov <destination>, (op) <source>

where:

<destination>	Is one of the destinations specified above.
<op>	If present, is:
	! or ~ for NOT (Note: this is always a bitwise NOT)
	:: for bit reverse
<source>	Is one of the sources specified above.

3.4.9. IRQ

3.4.9.1. Encoding

Bit:	15	14	13	12	11	10	9	8	7	6	5	4	3	2	1	0
IRQ	1	1	0	\multicolumn{5}{c}{Delay/side-set}			0	Clr	Wait	\multicolumn{4}{c}{Index}			

3.4.9.2. Operation

Set or clear the IRQ flag selected by `Index` argument.

- Clear: if 1, clear the flag selected by `Index`, instead of raising it. If `Clear` is set, the `Wait` bit has no effect.
- Wait: if 1, halt until the raised flag is lowered again, e.g. if a system interrupt handler has acknowledged the flag.
- Index:
    - The 3 LSBs specify an IRQ index from 0-7. This IRQ flag will be set/cleared depending on the Clear bit.
    - If the MSB is set, the state machine ID (0...3) is added to the IRQ index, by way of modulo-4 addition on the two LSBs. For example, state machine 2 with a flag value of 0x11 will raise flag 3, and a flag value of 0x13 will raise flag 1.

IRQ flags 4-7 are visible only to the state machines; IRQ flags 0-3 can be routed out to system level interrupts, on either of the PIO's two external interrupt request lines, configured by `IRQ0_INTE` and `IRQ1_INTE`.

The modulo addition bit allows relative addressing of 'IRQ' and 'WAIT' instructions, for synchronising state machines which are running the same program. Bit 2 (the third LSB) is unaffected by this addition.

If `Wait` is set, `Delay` cycles do not begin until after the wait period elapses.

3.4.9.3. Assembler Syntax

irq <irq_num> (rel)

irq set <irq_num> (rel)

irq nowait <irq_num> (rel)

irq wait <irq_num> (rel)

irq clear <irq_num> (rel)

where:

<irq_num> (rel)	Is a value (see Section 3.3.3) specifying The irq number to wait on (0-7). If *rel* is present, then the actual irq number used is calculating by replacing the low two bits of the irq number (*irq_num$_{10}$*) with the low two bits of the sum (*irq_num$_{10}$ + sm_num$_{10}$*) where *sm_num$_{10}$* is the state machine number
irq	Means set the IRQ without waiting
irq set	Also means set the IRQ without waiting
irq nowait	Again, means set the IRQ without waiting
irq wait	Means set the IRQ and wait for it to be cleared before proceeding
irq clear	Means clear the IRQ

3.4.10. SET

3.4.10.1. Encoding

Bit:	15	14	13	12	11	10	9	8	7	6	5	4	3	2	1	0
SET	1	1	1	\multicolumn{5}{c}{Delay/side-set}					\multicolumn{3}{c}{Destination}			\multicolumn{5}{c}{Data}				

3.4.10.2. Operation

Write immediate value `Data` to `Destination`.

- Destination:
    - 000: `PINS`
    - 001: `X` (scratch register X) 5 LSBs are set to `Data`, all others cleared to 0.
    - 010: `Y` (scratch register Y) 5 LSBs are set to `Data`, all others cleared to 0.
    - 011: Reserved
    - 100: `PINDIRS`
    - 101: Reserved
    - 110: Reserved
    - 111: Reserved
- Data: 5-bit immediate value to drive to pins or register.

This can be used to assert control signals such as a clock or chip select, or to initialise loop counters. As `Data` is 5 bits in size, scratch registers can be `SET` to values from 0-31, which is sufficient for a 32-iteration loop.

The mapping of `SET` and `OUT` onto pins is configured independently. They may be mapped to distinct locations, for example if one pin is to be used as a clock signal, and another for data. They may also be overlapping ranges of pins: a UART transmitter might use `SET` to assert start and stop bits, and `OUT` instructions to shift out FIFO data to the same pins.

3.4.10.3. Assembler Syntax

set <destination>, <value>

where:

<destination>   Is one of the destinations specified above.

<value>   The value (see Section 3.3.3) to set (valid range 0-31)

Chapter 4. Library documentation

Full library API documentation can be found online at https://raspberrypi.github.io/pico-sdk-doxygen/

Appendix A: App Notes

Attaching a 7 segment LED via GPIO

This example code shows how to interface the Raspberry Pi Pico to a generic 7 segment LED device. It uses the LED to count from 0 to 9 and then repeat. If the button is pressed, then the numbers will count down instead of up.

Wiring information

Our 7 Segment display has pins as follows.

```
  --A--
 F    B
  --G--
 E    C
  --D--
```

By default we are allocating GPIO 2 to segment A, 3 to B etc. So, connect GPIO 2 to pin A on the 7 segment LED display and so on. You will need the appropriate resistors (68 ohm should be fine) for each segment. The LED device used here is common anode, so the anode pin is connected to the 3.3v supply, and the GPIOs need to pull low (to ground) to complete the circuit. The pull direction of the GPIOs is specified in the code itself.

Connect the switch to connect on pressing. One side should be connected to ground, the other to GPIO 9.

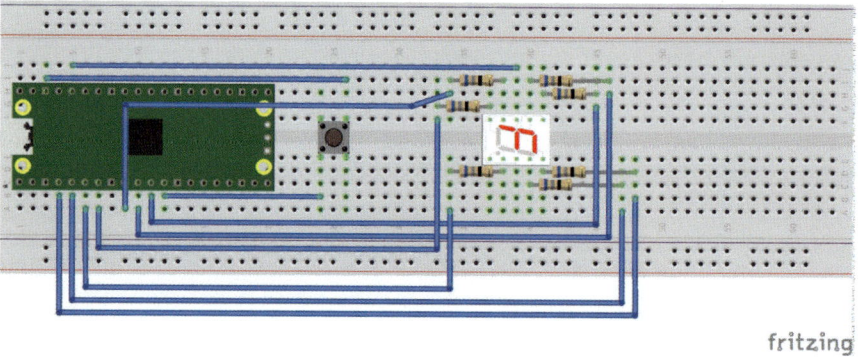

Figure 8. Wiring Diagram for 7 segment LED.

List of Files

CMakeLists.txt

CMake file to incorporate the example in to the examples build tree.

Pico Examples: *https://github.com/raspberrypi/pico-examples/blob/master/gpio/hello_7segment/CMakeLists.txt*

```
1  add_executable(hello_7segment
2          hello_7segment.c
3          )
4
5  # pull in common dependencies
6  target_link_libraries(hello_7segment pico_stdlib)
7
8  # create map/bin/hex file etc.
```

```
 9  pico_add_extra_outputs(hello_7segment)
10
11  # add url via pico_set_program_url
12  example_auto_set_url(hello_7segment)
```

hello_7segment.c

The example code.

Pico Examples: *https://github.com/raspberrypi/pico-examples/blob/master/gpio/hello_7segment/hello_7segment.c*

```c
 1  /**
 2   * Copyright (c) 2020 Raspberry Pi (Trading) Ltd.
 3   *
 4   * SPDX-License-Identifier: BSD-3-Clause
 5   */
 6
 7  #include <stdio.h>
 8  #include "pico/stdlib.h"
 9  #include "hardware/gpio.h"
10
11  /*
12    Our 7 Segment display has pins as follows:
13
14    --A--
15    F   B
16    --G--
17    E   C
18    --D--
19
20    By default we are allocating GPIO 2 to segment A, 3 to B etc.
21    So, connect GPIO 2 to pin A on the 7 segment LED display etc. Don't forget
22    the appropriate resistors, best to use one for each segment!
23
24    Connect button so that pressing the switch connects the GPIO 9 (default) to
25    ground (pull down)
26  */
27
28  #define FIRST_GPIO 2
29  #define BUTTON_GPIO (FIRST_GPIO+7)
30
31  // This array converts a number 0-9 to a bit pattern to send to the GPIOs
32  int bits[10] = {
33          0x3f,  // 0
34          0x06,  // 1
35          0x5b,  // 2
36          0x4f,  // 3
37          0x66,  // 4
38          0x6d,  // 5
39          0x7d,  // 6
40          0x07,  // 7
41          0x7f,  // 8
42          0x67   // 9
43  };
44
45  /// \tag::hello_gpio[]
46  int main() {
47      stdio_init_all();
48      printf("Hello, 7segment - press button to count down!\n");
49
50      // We could use gpio_set_dir_out_masked() here
51      for (int gpio = FIRST_GPIO; gpio < FIRST_GPIO + 7; gpio++) {
```

```c
52          gpio_init(gpio);
53          gpio_set_dir(gpio, GPIO_OUT);
54          // Our bitmap above has a bit set where we need an LED on, BUT, we are pulling low to light
55          // so invert our output
56          gpio_set_outover(gpio, GPIO_OVERRIDE_INVERT);
57      }
58
59      gpio_init(BUTTON_GPIO);
60      gpio_set_dir(BUTTON_GPIO, GPIO_IN);
61      // We are using the button to pull down to 0v when pressed, so ensure that when
62      // unpressed, it uses internal pull ups. Otherwise when unpressed, the input will
63      // be floating.
64      gpio_pull_up(BUTTON_GPIO);
65
66      int val = 0;
67      while (true) {
68          // Count upwards or downwards depending on button input
69          // We are pulling down on switch active, so invert the get to make
70          // a press count downwards
71          if (!gpio_get(BUTTON_GPIO)) {
72              if (val == 9) {
73                  val = 0;
74              } else {
75                  val++;
76              }
77          } else if (val == 0) {
78              val = 9;
79          } else {
80              val--;
81          }
82
83          // We are starting with GPIO 2, our bitmap starts at bit 0 so shift to start at 2.
84          int32_t mask = bits[val] << FIRST_GPIO;
85
86          // Set all our GPIOs in one go!
87          // If something else is using GPIO, we might want to use gpio_put_masked()
88          gpio_set_mask(mask);
89          sleep_ms(250);
90          gpio_clr_mask(mask);
91      }
92
93      return 0;
94  }
95  /// \end::hello_gpio[]
```

Bill of Materials

Table 9. A list of materials required for the example

Item	Quantity	Details
Breadboard	1	generic part
Raspberry Pi Pico	1	https://www.raspberrypi.com/products/raspberry-pi-pico/
7 segment LED module	1	generic part
68 ohm resistor	7	generic part
DIL push to make switch	1	generic switch

M/M Jumper wires	10	generic part

DHT-11, DHT-22, and AM2302 Sensors

The DHT sensors are fairly well known hobbyist sensors for measuring relative humidity and temperature using a capacitive humidity sensor, and a thermistor. While they are slow, one reading every ~2 seconds, they are reliable and good for basic data logging. Communication is based on a custom protocol which uses a single wire for data.

> **ℹ NOTE**
>
> The DHT-11 and DHT-22 sensors are the most common. They use the same protocol but have different characteristics, the DHT-22 has better accuracy, and has a larger sensor range than the DHT-11. The sensor is available from a number of retailers.

Wiring information

See Figure 9 for wiring instructions.

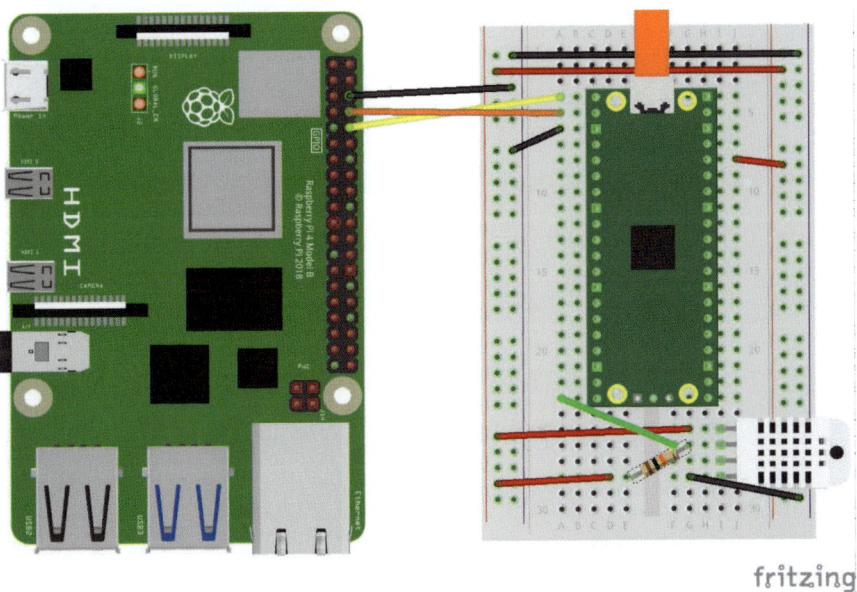

Figure 9. Wiring the DHT-22 temperature sensor to Raspberry Pi Pico, and connecting Pico's UART0 to the Raspberry Pi 4.

> **ℹ NOTE**
>
> One of the pins (pin 3) on the DHT sensor will not be connected, it is not used.

You will want to place a 10 kΩ resistor between VCC and the data pin, to act as a medium-strength pull up on the data line.

Connecting UART0 of Pico to Raspberry Pi as in Figure 9 and you should see something similar to Figure 10 in `minicom` when connected to `/dev/serial0` on the Raspberry Pi.

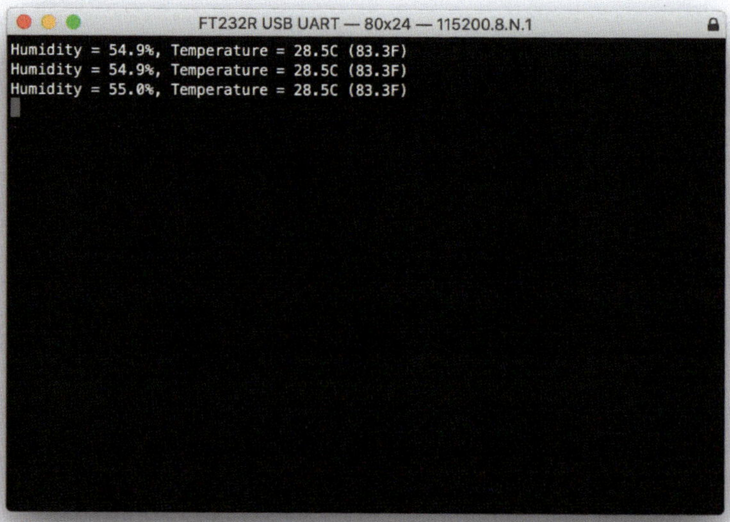

Figure 10. Serial output over Pico's UART0 in a terminal window.

```
Humidity = 54.9%, Temperature = 28.5C (83.3F)
Humidity = 54.9%, Temperature = 28.5C (83.3F)
Humidity = 55.0%, Temperature = 28.5C (83.3F)
```

Connect to `/dev/serial0` by typing,

```
$ minicom -b 115200 -o -D /dev/serial0
```

at the command line.

List of Files

A list of files with descriptions of their function;

CMakeLists.txt

Make file to incorporate the example in to the examples build tree.

Pico Examples: https://github.com/raspberrypi/pico-examples/blob/master/gpio/dht_sensor/CMakeLists.txt

```
1  add_executable(dht
2          dht.c
3          )
4
5  target_link_libraries(dht pico_stdlib)
6
7  pico_add_extra_outputs(dht)
8
9  # add url via pico_set_program_url
10 example_auto_set_url(dht)
```

dht.c

The example code.

Pico Examples: https://github.com/raspberrypi/pico-examples/blob/master/gpio/dht_sensor/dht.c

```
1  /**
2   * Copyright (c) 2020 Raspberry Pi (Trading) Ltd.
3   *
```

```c
 4  * SPDX-License-Identifier: BSD-3-Clause
 5  **/
 6
 7 #include <stdio.h>
 8 #include <math.h>
 9 #include "pico/stdlib.h"
10 #include "hardware/gpio.h"
11
12 #ifdef PICO_DEFAULT_LED_PIN
13 #define LED_PIN PICO_DEFAULT_LED_PIN
14 #endif
15
16 const uint DHT_PIN = 15;
17 const uint MAX_TIMINGS = 85;
18
19 typedef struct {
20     float humidity;
21     float temp_celsius;
22 } dht_reading;
23
24 void read_from_dht(dht_reading *result);
25
26 int main() {
27     stdio_init_all();
28     gpio_init(DHT_PIN);
29 #ifdef LED_PIN
30     gpio_init(LED_PIN);
31     gpio_set_dir(LED_PIN, GPIO_OUT);
32 #endif
33     while (1) {
34         dht_reading reading;
35         read_from_dht(&reading);
36         float fahrenheit = (reading.temp_celsius * 9 / 5) + 32;
37         printf("Humidity = %.1f%%, Temperature = %.1fC (%.1fF)\n",
38                 reading.humidity, reading.temp_celsius, fahrenheit);
39
40         sleep_ms(2000);
41     }
42 }
43
44 void read_from_dht(dht_reading *result) {
45     int data[5] = {0, 0, 0, 0, 0};
46     uint last = 1;
47     uint j = 0;
48
49     gpio_set_dir(DHT_PIN, GPIO_OUT);
50     gpio_put(DHT_PIN, 0);
51     sleep_ms(20);
52     gpio_set_dir(DHT_PIN, GPIO_IN);
53
54 #ifdef LED_PIN
55     gpio_put(LED_PIN, 1);
56 #endif
57     for (uint i = 0; i < MAX_TIMINGS; i++) {
58         uint count = 0;
59         while (gpio_get(DHT_PIN) == last) {
60             count++;
61             sleep_us(1);
62             if (count == 255) break;
63         }
64         last = gpio_get(DHT_PIN);
65         if (count == 255) break;
66
```

```
67          if ((i >= 4) && (i % 2 == 0)) {
68              data[j / 8] <<= 1;
69              if (count > 16) data[j / 8] |= 1;
70              j++;
71          }
72      }
73  #ifdef LED_PIN
74      gpio_put(LED_PIN, 0);
75  #endif
76
77      if ((j >= 40) && (data[4] == ((data[0] + data[1] + data[2] + data[3]) & 0xFF))) {
78          result->humidity = (float) ((data[0] << 8) + data[1]) / 10;
79          if (result->humidity > 100) {
80              result->humidity = data[0];
81          }
82          result->temp_celsius = (float) (((data[2] & 0x7F) << 8) + data[3]) / 10;
83          if (result->temp_celsius > 125) {
84              result->temp_celsius = data[2];
85          }
86          if (data[2] & 0x80) {
87              result->temp_celsius = -result->temp_celsius;
88          }
89      } else {
90          printf("Bad data\n");
91      }
92  }
```

Bill of Materials

Table 10. A list of materials required for the example

Item	Quantity	Details
Breadboard	1	generic part
Raspberry Pi Pico	1	https://www.raspberrypi.com/products/raspberry-pi-pico/
10 kΩ resistor	1	generic part
M/M Jumper wires	4	generic part
DHT-22 sensor	1	generic part

Attaching a 16x2 LCD via TTL

This example code shows how to interface the Raspberry Pi Pico to one of the very common 16x2 LCD character displays. Due to the large number of pins these displays use, they are commonly used with extra drivers or backpacks. In this example, we will use an Adafruit LCD display backpack, which supports communication over USB or TTL. A monochrome display with an RGB backlight is also used, but the backpack is compatible with monochrome backlight displays too. There is another example that uses I2C to control a 16x2 display.

The backpack processes a set of commands that are documented here and preceded by the "special" byte 0xFE. The backpack does the ASCII character conversion and even supports custom character creation. In this example, we use the Pico's primary UART (uart0) to read characters from our computer and send them via the other UART (uart1) to print them onto the LCD. We also define a special startup sequence and vary the display's backlight color.

> **NOTE**
>
> You can change where stdio output goes (Pico's USB, uart0 or both) with CMake directives. The CMakeLists.txt file shows how to enable both.

Wiring information

Wiring up the backpack to the Pico requires 3 jumpers, to connect VCC (3.3v), GND, TX. The example here uses both of the Pico's UARTs, one (uart0) for stdio and the other (uart1) for communication with the backpack. Pin 8 is used as the TX pin. Power is supplied from the 3.3V pin. To connect the backpack to the display, it is common practice to solder it onto the back of the display, or during the prototyping stage to use the same parallel lanes on a breadboard.

> **NOTE**
>
> While this display will work at 3.3V, it will be quite dim. Using a 5V source will make it brighter.

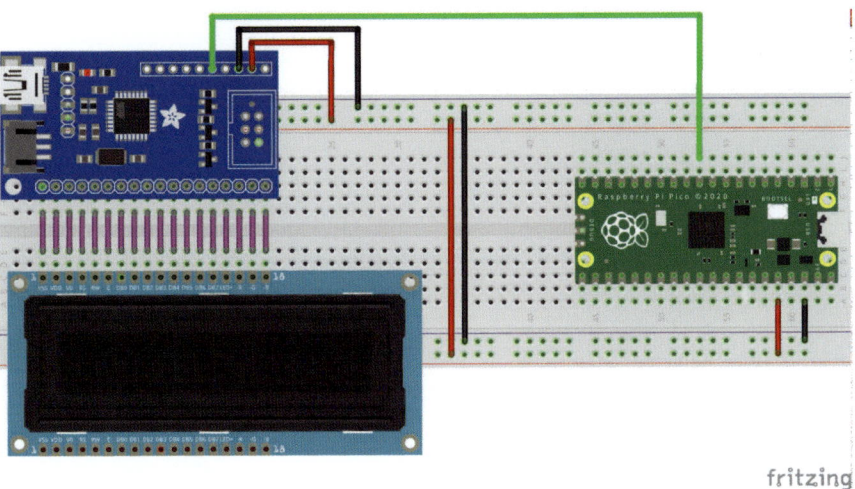

Figure 11. Wiring Diagram for LCD with TTL backpack.

List of Files

CMakeLists.txt

CMake file to incorporate the example in to the examples build tree.

Pico Examples: https://github.com/raspberrypi/pico-examples/blob/master/uart/lcd_uart/CMakeLists.txt

```
1  add_executable(lcd_uart
2          lcd_uart.c
3          )
4
5  # pull in common dependencies and additional uart hardware support
6  target_link_libraries(lcd_uart pico_stdlib hardware_uart)
7
8  # enable usb output and uart output
9  # modify here as required
10 pico_enable_stdio_usb(lcd_uart 1)
11 pico_enable_stdio_uart(lcd_uart 1)
12
13 # create map/bin/hex file etc.
14 pico_add_extra_outputs(lcd_uart)
15
16 # add url via pico_set_program_url
```

```
17 example_auto_set_url(lcd_uart)
```

lcd_uart.c

The example code.

Pico Examples: *https://github.com/raspberrypi/pico-examples/blob/master/uart/lcd_uart/lcd_uart.c*

```c
/**
 * Copyright (c) 2021 Raspberry Pi (Trading) Ltd.
 *
 * SPDX-License-Identifier: BSD-3-Clause
 */

/* Example code to drive a 16x2 LCD panel via an Adafruit TTL LCD "backpack"

   Optionally, the backpack can be connected the VBUS (pin 40) at 5V if
   the Pico in question is powered by USB for greater brightness.

   If this is done, then no other connections should be made to the backpack apart
   from those listed below as the backpack's logic levels will change.

   Connections on Raspberry Pi Pico board, other boards may vary.

   GPIO 8 (pin 11)-> RX on backpack
   3.3v (pin 36) -> 3.3v on backpack
   GND (pin 38)  -> GND on backpack
*/

#include <stdio.h>
#include <math.h>
#include "pico/stdlib.h"
#include "pico/binary_info.h"
#include "hardware/uart.h"

 // leave uart0 free for stdio
#define UART_ID uart1
#define BAUD_RATE 9600
#define UART_TX_PIN 8
#define LCD_WIDTH 16
#define LCD_HEIGHT 2

// basic commands
#define LCD_DISPLAY_ON 0x42
#define LCD_DISPLAY_OFF 0x46
#define LCD_SET_BRIGHTNESS 0x99
#define LCD_SET_CONTRAST 0x50
#define LCD_AUTOSCROLL_ON 0x51
#define LCD_AUTOSCROLL_OFF 0x52
#define LCD_CLEAR_SCREEN 0x58
#define LCD_SET_SPLASH 0x40

// cursor commands
#define LCD_SET_CURSOR_POS 0x47
#define LCD_CURSOR_HOME 0x48
#define LCD_CURSOR_BACK 0x4C
#define LCD_CURSOR_FORWARD 0x4D
#define LCD_UNDERLINE_CURSOR_ON 0x4A
#define LCD_UNDERLINE_CURSOR_OFF 0x4B
#define LCD_BLOCK_CURSOR_ON 0x53
#define LCD_BLOCK_CURSOR_OFF 0x54

```

```c
55  // rgb commands
56  #define LCD_SET_BACKLIGHT_COLOR 0xD0
57  #define LCD_SET_DISPLAY_SIZE 0xD1
58
59  // change to 0 if display is not RGB capable
60  #define LCD_IS_RGB 1
61
62  void lcd_write(uint8_t cmd, uint8_t* buf, uint8_t buflen) {
63      // all commands are prefixed with 0xFE
64      const uint8_t pre = 0xFE;
65      uart_write_blocking(UART_ID, &pre, 1);
66      uart_write_blocking(UART_ID, &cmd, 1);
67      uart_write_blocking(UART_ID, buf, buflen);
68      sleep_ms(10); // give the display some time
69  }
70
71  void lcd_set_size(uint8_t w, uint8_t h) {
72      // sets the dimensions of the display
73      uint8_t buf[] = { w, h };
74      lcd_write(LCD_SET_DISPLAY_SIZE, buf, 2);
75  }
76
77  void lcd_set_contrast(uint8_t contrast) {
78      // sets the display contrast
79      lcd_write(LCD_SET_CONTRAST, &contrast, 1);
80  }
81
82  void lcd_set_brightness(uint8_t brightness) {
83      // sets the backlight brightness
84      lcd_write(LCD_SET_BRIGHTNESS, &brightness, 1);
85  }
86
87  void lcd_set_cursor(bool is_on) {
88      // set is_on to true if we want the blinking block and underline cursor to show
89      if (is_on) {
90          lcd_write(LCD_BLOCK_CURSOR_ON, NULL, 0);
91          lcd_write(LCD_UNDERLINE_CURSOR_ON, NULL, 0);
92      } else {
93          lcd_write(LCD_BLOCK_CURSOR_OFF, NULL, 0);
94          lcd_write(LCD_UNDERLINE_CURSOR_OFF, NULL, 0);
95      }
96  }
97
98  void lcd_set_backlight(bool is_on) {
99      // turn the backlight on (true) or off (false)
100     if (is_on) {
101         lcd_write(LCD_DISPLAY_ON, (uint8_t *) 0, 1);
102     } else {
103         lcd_write(LCD_DISPLAY_OFF, NULL, 0);
104     }
105 }
106
107 void lcd_clear() {
108     // clear the contents of the display
109     lcd_write(LCD_CLEAR_SCREEN, NULL, 0);
110 }
111
112 void lcd_cursor_reset() {
113     // reset the cursor to (1, 1)
114     lcd_write(LCD_CURSOR_HOME, NULL, 0);
115 }
116
117 #if LCD_IS_RGB
```

```c
118  void lcd_set_backlight_color(uint8_t r, uint8_t g, uint8_t b) {
119      // only supported on RGB displays!
120      uint8_t buf[] = { r, g, b };
121      lcd_write(LCD_SET_BACKLIGHT_COLOR, buf, 3);
122  }
123  #endif
124
125  void lcd_init() {
126      lcd_set_backlight(true);
127      lcd_set_size(LCD_WIDTH, LCD_HEIGHT);
128      lcd_set_contrast(155);
129      lcd_set_brightness(255);
130      lcd_set_cursor(false);
131  }
132
133  int main() {
134      stdio_init_all();
135      uart_init(UART_ID, BAUD_RATE);
136      uart_set_translate_crlf(UART_ID, false);
137      gpio_set_function(UART_TX_PIN, GPIO_FUNC_UART);
138
139      bi_decl(bi_1pin_with_func(UART_TX_PIN, GPIO_FUNC_UART));
140
141      lcd_init();
142
143      // define startup sequence and save to EEPROM
144      // no more or less than 32 chars, if not enough, fill remaining ones with spaces
145      uint8_t splash_buf[] = "Hello LCD, from Pi Towers!       ";
146      lcd_write(LCD_SET_SPLASH, splash_buf, LCD_WIDTH * LCD_HEIGHT);
147
148      lcd_cursor_reset();
149      lcd_clear();
150
151  #if LCD_IS_RGB
152      uint8_t i = 0; // it's ok if this overflows and wraps, we're using sin
153      const float frequency = 0.1f;
154      float red, green, blue;
155  #endif
156
157      while (1) {
158          // send any chars from stdio straight to the backpack
159          char c = getchar();
160          // any bytes not followed by 0xFE (the special command) are interpreted
161          // as text to be displayed on the backpack, so we just send the char
162          // down the UART byte pipe!
163          if (c < 128) uart_putc_raw(UART_ID, c); // skip extra non-ASCII chars
164  #if LCD_IS_RGB
165          // change the display color on keypress, rainbow style!
166          red = sin(frequency * i + 0) * 127 + 128;
167          green = sin(frequency * i + 2) * 127 + 128;
168          blue = sin(frequency * i + 4) * 127 + 128;
169          lcd_set_backlight_color(red, green, blue);
170          i++;
171  #endif
172      }
173  }
```

Bill of Materials

Table 11. A list of materials required for the example

Item	Quantity	Details
Breadboard	1	generic part
Raspberry Pi Pico	1	https://www.raspberrypi.com/products/raspberry-pi-pico/
16x2 RGB LCD panel 3.3v	1	generic part, available on Adafruit
16x2 LCD backpack	1	from Adafruit
M/M Jumper wires	3	generic part

Attaching a microphone using the ADC

This example code shows how to interface the Raspberry Pi Pico with a standard analog microphone via the onboard analog to digital converter (ADC). In this example, we use an ICS-40180 breakout board by SparkFun but any analog microphone should be compatible with this tutorial. SparkFun have written a guide for this board that goes into more detail about the board and how it works.

> **TIP**
>
> An analog to digital converter (ADC) is responsible for reading continually varying input signals that may range from 0 to a specified reference voltage (in the Pico's case this reference voltage is set by the supply voltage and can be measured on pin 35, ADC_VREF) and converting them into binary, i.e. a number that can be digitally stored.

The Pico has a 12-bit ADC (ENOB of 8.7-bit, see RP2040 datasheet section 4.9.3 for more details), meaning that a read operation will return a number ranging from 0 to 4095 (2^12 - 1) for a total of 4096 possible values. Therefore, the resolution of the ADC is 3.3/4096, so roughly steps of 0.8 millivolts. The SparkFun breakout uses an OPA344 operational amplifier to boost the signal coming from the microphone to voltage levels that can be easily read by the ADC. An important side effect is that a bias of 0.5*Vcc is added to the signal, even when the microphone is not picking up any sound.

The ADC provides us with a raw voltage value but when dealing with sound, we're more interested in the amplitude of the audio signal. This is defined as one half the peak-to-peak amplitude. Included with this example is a very simple Python script that will plot the voltage values it receives via the serial port. By tweaking the sampling rates, and various other parameters, the data from the microphone can be analysed in various ways, such as in a Fast Fourier Transform to see what frequencies make up the signal.

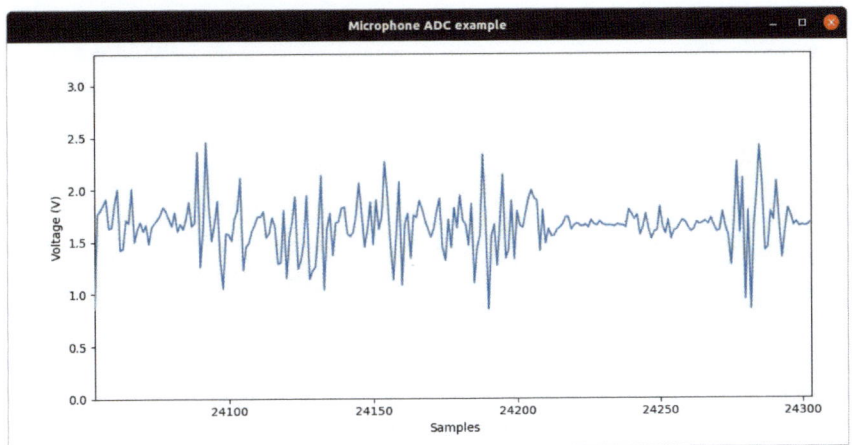

Figure 12. Example output from included Python script

Wiring information

Wiring up the device requires 3 jumpers, to connect VCC (3.3v), GND, and AOUT. The example here uses ADC0, which is GP26. Power is supplied from the 3.3V pin.

> ⛔ **WARNING**
>
> Most boards will take a range of VCC voltages from the Pico's default 3.3V to the 5 volts commonly seen on other microcontrollers. Ensure your board doesn't output an analogue signal greater than 3.3V as this may result in permanent damage to the Pico's ADC.

Figure 13. Wiring Diagram for ICS-40180 microphone breakout board.

List of Files

CMakeLists.txt

    CMake file to incorporate the example in to the examples build tree.

Pico Examples: *https://github.com/raspberrypi/pico-examples/blob/master/adc/microphone_adc/CMakeLists.txt*

```
1  add_executable(microphone_adc
2          microphone_adc.c
3          )
4
5  # pull in common dependencies and adc hardware support
6  target_link_libraries(microphone_adc pico_stdlib hardware_adc)
7
8  # create map/bin/hex file etc.
9  pico_add_extra_outputs(microphone_adc)
10
11 # add url via pico_set_program_url
12 example_auto_set_url(microphone_adc)
```

microphone_adc.c

    The example code.

Pico Examples: *https://github.com/raspberrypi/pico-examples/blob/master/adc/microphone_adc/microphone_adc.c*

```c
/**
 * Copyright (c) 2021 Raspberry Pi (Trading) Ltd.
 *
 * SPDX-License-Identifier: BSD-3-Clause
 */

#include <stdio.h>
#include "pico/stdlib.h"
#include "hardware/gpio.h"
#include "hardware/adc.h"
#include "hardware/uart.h"
#include "pico/binary_info.h"

/* Example code to extract analog values from a microphone using the ADC
   with accompanying Python file to plot these values

   Connections on Raspberry Pi Pico board, other boards may vary.

   GPIO 26/ADC0 (pin 31)-> AOUT or AUD on microphone board
   3.3v (pin 36)  -> VCC on microphone board
   GND (pin 38)   -> GND on microphone board
*/

#define ADC_NUM 0
#define ADC_PIN (26 + ADC_NUM)
#define ADC_VREF 3.3
#define ADC_RANGE (1 << 12)
#define ADC_CONVERT (ADC_VREF / (ADC_RANGE - 1))

int main() {
    stdio_init_all();
    printf("Beep boop, listening...\n");

    bi_decl(bi_program_description("Analog microphone example for Raspberry Pi Pico")); // for picotool
    bi_decl(bi_1pin_with_name(ADC_PIN, "ADC input pin"));

    adc_init();
    adc_gpio_init( ADC_PIN);
    adc_select_input( ADC_NUM);

    uint adc_raw;
    while (1) {
        adc_raw = adc_read(); // raw voltage from ADC
        printf("%.2f\n", adc_raw * ADC_CONVERT);
        sleep_ms(10);
    }

    return 0;
}
```

Bill of Materials

Table 12. A list of materials required for the example

Item	Quantity	Details
Breadboard	1	generic part

Raspberry Pi Pico	1	https://www.raspberrypi.com/products/raspberry-pi-pico/
ICS-40180 microphone breakout board or similar	1	From SparkFun
M/M Jumper wires	3	generic part

Attaching a BME280 temperature/humidity/pressure sensor via SPI

This example code shows how to interface the Raspberry Pi Pico to a BME280 temperature/humidity/pressure. The particular device used can be interfaced via I2C or SPI, we are using SPI, and interfacing at 3.3v.

This examples reads the data from the sensor, and runs it through the appropriate compensation routines (see the chip datasheet for details https://www.bosch-sensortec.com/media/boschsensortec/downloads/datasheets/bst-bme280-ds002.pdf). At startup the compensation parameters required by the compensation routines are read from the chip.)

Wiring information

Wiring up the device requires 6 jumpers as follows:

- GPIO 16 (pin 21) MISO/spi0_rx → SDO/SDO on bme280 board
- GPIO 17 (pin 22) Chip select → CSB/!CS on bme280 board
- GPIO 18 (pin 24) SCK/spi0_sclk → SCL/SCK on bme280 board
- GPIO 19 (pin 25) MOSI/spi0_tx → SDA/SDI on bme280 board
- 3.3v (pin 3;6) → VCC on bme280 board
- GND (pin 38) → GND on bme280 board

The example here uses SPI port 0. Power is supplied from the 3.3V pin.

> **NOTE**
>
> There are many different manufacturers who sell boards with the BME280. Whilst they all appear slightly different, they all have, at least, the same 6 pins required to power and communicate. When wiring up a board that is different to the one in the diagram, ensure you connect up as described in the previous paragraph.

Figure 14. Wiring Diagram for bme280.

List of Files

CMakeLists.txt

CMake file to incorporate the example in to the examples build tree.

Pico Examples: https://github.com/raspberrypi/pico-examples/blob/master/spi/bme280_spi/CMakeLists.txt

```cmake
add_executable(bme280_spi
        bme280_spi.c
        )

# pull in common dependencies and additional spi hardware support
target_link_libraries(bme280_spi pico_stdlib hardware_spi)

# create map/bin/hex file etc.
pico_add_extra_outputs(bme280_spi)

# add url via pico_set_program_url
example_auto_set_url(bme280_spi)
```

bme280_spi.c

The example code.

Pico Examples: https://github.com/raspberrypi/pico-examples/blob/master/spi/bme280_spi/bme280_spi.c

```c
/**
 * Copyright (c) 2020 Raspberry Pi (Trading) Ltd.
 *
 * SPDX-License-Identifier: BSD-3-Clause
 */

#include <stdio.h>
#include <string.h>
#include "pico/stdlib.h"
#include "pico/binary_info.h"
#include "hardware/spi.h"

/* Example code to talk to a bme280 humidity/temperature/pressure sensor.

   NOTE: Ensure the device is capable of being driven at 3.3v NOT 5v. The Pico
   GPIO (and therefore SPI) cannot be used at 5v.

   You will need to use a level shifter on the SPI lines if you want to run the
   board at 5v.

   Connections on Raspberry Pi Pico board and a generic bme280 board, other
   boards may vary.

   GPIO 16 (pin 21) MISO/spi0_rx-> SDO/SDO on bme280 board
   GPIO 17 (pin 22) Chip select -> CSB/!CS on bme280 board
   GPIO 18 (pin 24) SCK/spi0_sclk -> SCL/SCK on bme280 board
   GPIO 19 (pin 25) MOSI/spi0_tx -> SDA/SDI on bme280 board
   3.3v (pin 36) -> VCC on bme280 board
   GND (pin 38)  -> GND on bme280 board

   Note: SPI devices can have a number of different naming schemes for pins. See
   the Wikipedia page at https://en.wikipedia.org/wiki/Serial_Peripheral_Interface
   for variations.

   This code uses a bunch of register definitions, and some compensation code derived
   from the Bosch datasheet which can be found here.
   https://www.bosch-sensortec.com/media/boschsensortec/downloads/datasheets/bst-bme280-
```

```
       ds002.pdf
38  */
39
40  #define READ_BIT 0x80
41
42  int32_t t_fine;
43
44  uint16_t dig_T1;
45  int16_t dig_T2, dig_T3;
46  uint16_t dig_P1;
47  int16_t dig_P2, dig_P3, dig_P4, dig_P5, dig_P6, dig_P7, dig_P8, dig_P9;
48  uint8_t dig_H1, dig_H3;
49  int8_t dig_H6;
50  int16_t dig_H2, dig_H4, dig_H5;
51
52  /* The following compensation functions are required to convert from the raw ADC
53  data from the chip to something usable. Each chip has a different set of
54  compensation parameters stored on the chip at point of manufacture, which are
55  read from the chip at startup and used in these routines.
56  */
57  int32_t compensate_temp(int32_t adc_T) {
58      int32_t var1, var2, T;
59      var1 = ((((adc_T >> 3) - ((int32_t) dig_T1 << 1))) * ((int32_t) dig_T2)) >> 11;
60      var2 = (((((adc_T >> 4) - ((int32_t) dig_T1)) * ((adc_T >> 4) - ((int32_t) dig_T1))) >> 12) * ((int32_t) dig_T3))
61              >> 14;
62
63      t_fine = var1 + var2;
64      T = (t_fine * 5 + 128) >> 8;
65      return T;
66  }
67
68  uint32_t compensate_pressure(int32_t adc_P) {
69      int32_t var1, var2;
70      uint32_t p;
71      var1 = (((int32_t) t_fine) >> 1) - (int32_t) 64000;
72      var2 = (((var1 >> 2) * (var1 >> 2)) >> 11) * ((int32_t) dig_P6);
73      var2 = var2 + ((var1 * ((int32_t) dig_P5)) << 1);
74      var2 = (var2 >> 2) + (((int32_t) dig_P4) << 16);
75      var1 = (((dig_P3 * (((var1 >> 2) * (var1 >> 2)) >> 13)) >> 3) + ((((int32_t) dig_P2) * var1) >> 1)) >> 18;
76      var1 = ((((32768 + var1)) * ((int32_t) dig_P1)) >> 15);
77      if (var1 == 0)
78          return 0;
79
80      p = (((uint32_t) (((int32_t) 1048576) - adc_P) - (var2 >> 12))) * 3125;
81      if (p < 0x80000000)
82          p = (p << 1) / ((uint32_t) var1);
83      else
84          p = (p / (uint32_t) var1) * 2;
85
86      var1 = (((int32_t) dig_P9) * ((int32_t) (((p >> 3) * (p >> 3)) >> 13))) >> 12;
87      var2 = (((int32_t) (p >> 2)) * ((int32_t) dig_P8)) >> 13;
88      p = (uint32_t) ((int32_t) p + ((var1 + var2 + dig_P7) >> 4));
89
90      return p;
91  }
92
93  uint32_t compensate_humidity(int32_t adc_H) {
94      int32_t v_x1_u32r;
95      v_x1_u32r = (t_fine - ((int32_t) 76800));
96      v_x1_u32r = (((((adc_H << 14) - (((int32_t) dig_H4) << 20) - (((int32_t) dig_H5) * v_x1_u32r)) +
```

```
 97                        ((int32_t) 16384)) >> 15) * (((((((v_x1_u32r * ((int32_t) dig_H6)) >>
       10) * (((v_x1_u32r *
 98
       ((int32_t) dig_H3))
 99              >> 11) + ((int32_t) 32768))) >> 10) + ((int32_t) 2097152)) *
100                                                ((int32_t) dig_H2) + 8192) >> 14));
101     v_x1_u32r = (v_x1_u32r - (((((v_x1_u32r >> 15) * (v_x1_u32r >> 15)) >> 7) * ((int32_t)
       dig_H1)) >> 4));
102     v_x1_u32r = (v_x1_u32r < 0 ? 0 : v_x1_u32r);
103     v_x1_u32r = (v_x1_u32r > 419430400 ? 419430400 : v_x1_u32r);
104
105     return (uint32_t) (v_x1_u32r >> 12);
106 }
107
108 #ifdef PICO_DEFAULT_SPI_CSN_PIN
109 static inline void cs_select() {
110     asm volatile("nop \n nop \n nop");
111     gpio_put(PICO_DEFAULT_SPI_CSN_PIN, 0);  // Active low
112     asm volatile("nop \n nop \n nop");
113 }
114
115 static inline void cs_deselect() {
116     asm volatile("nop \n nop \n nop");
117     gpio_put(PICO_DEFAULT_SPI_CSN_PIN, 1);
118     asm volatile("nop \n nop \n nop");
119 }
120 #endif
121
122 #if defined(spi_default) && defined(PICO_DEFAULT_SPI_CSN_PIN)
123 static void write_register(uint8_t reg, uint8_t data) {
124     uint8_t buf[2];
125     buf[0] = reg & 0x7f;  // remove read bit as this is a write
126     buf[1] = data;
127     cs_select();
128     spi_write_blocking(spi_default, buf, 2);
129     cs_deselect();
130     sleep_ms(10);
131 }
132
133 static void read_registers(uint8_t reg, uint8_t *buf, uint16_t len) {
134     // For this particular device, we send the device the register we want to read
135     // first, then subsequently read from the device. The register is auto incrementing
136     // so we don't need to keep sending the register we want, just the first.
137     reg |= READ_BIT;
138     cs_select();
139     spi_write_blocking(spi_default, &reg, 1);
140     sleep_ms(10);
141     spi_read_blocking(spi_default, 0, buf, len);
142     cs_deselect();
143     sleep_ms(10);
144 }
145
146 /* This function reads the manufacturing assigned compensation parameters from the device */
147 void read_compensation_parameters() {
148     uint8_t buffer[26];
149
150     read_registers(0x88, buffer, 24);
151
152     dig_T1 = buffer[0] | (buffer[1] << 8);
153     dig_T2 = buffer[2] | (buffer[3] << 8);
154     dig_T3 = buffer[4] | (buffer[5] << 8);
155
156     dig_P1 = buffer[6] | (buffer[7] << 8);
```

```c
157         dig_P2 = buffer[8] | (buffer[9] << 8);
158         dig_P3 = buffer[10] | (buffer[11] << 8);
159         dig_P4 = buffer[12] | (buffer[13] << 8);
160         dig_P5 = buffer[14] | (buffer[15] << 8);
161         dig_P6 = buffer[16] | (buffer[17] << 8);
162         dig_P7 = buffer[18] | (buffer[19] << 8);
163         dig_P8 = buffer[20] | (buffer[21] << 8);
164         dig_P9 = buffer[22] | (buffer[23] << 8);
165
166         dig_H1 = buffer[25];
167
168         read_registers(0xE1, buffer, 8);
169
170         dig_H2 = buffer[0] | (buffer[1] << 8);
171         dig_H3 = (int8_t) buffer[2];
172         dig_H4 = buffer[3] << 4 | (buffer[4] & 0xf);
173         dig_H5 = (buffer[5] >> 4) | (buffer[6] << 4);
174         dig_H6 = (int8_t) buffer[7];
175 }
176
177 static void bme280_read_raw(int32_t *humidity, int32_t *pressure, int32_t *temperature) {
178         uint8_t buffer[8];
179
180         read_registers(0xF7, buffer, 8);
181         *pressure = ((uint32_t) buffer[0] << 12) | ((uint32_t) buffer[1] << 4) | (buffer[2] >> 4);
182         *temperature = ((uint32_t) buffer[3] << 12) | ((uint32_t) buffer[4] << 4) | (buffer[5] >> 4);
183         *humidity = (uint32_t) buffer[6] << 8 | buffer[7];
184 }
185 #endif
186
187 int main() {
188         stdio_init_all();
189 #if !defined(spi_default) || !defined(PICO_DEFAULT_SPI_SCK_PIN) || !defined(PICO_DEFAULT_SPI_TX_PIN) || !defined(PICO_DEFAULT_SPI_RX_PIN) || !defined(PICO_DEFAULT_SPI_CSN_PIN)
190 #warning spi/bme280_spi example requires a board with SPI pins
191         puts("Default SPI pins were not defined");
192 #else
193
194         printf("Hello, bme280! Reading raw data from registers via SPI...\n");
195
196         // This example will use SPI0 at 0.5MHz.
197         spi_init(spi_default, 500 * 1000);
198         gpio_set_function(PICO_DEFAULT_SPI_RX_PIN, GPIO_FUNC_SPI);
199         gpio_set_function(PICO_DEFAULT_SPI_SCK_PIN, GPIO_FUNC_SPI);
200         gpio_set_function(PICO_DEFAULT_SPI_TX_PIN, GPIO_FUNC_SPI);
201         // Make the SPI pins available to picotool
202         bi_decl(bi_3pins_with_func(PICO_DEFAULT_SPI_RX_PIN, PICO_DEFAULT_SPI_TX_PIN, PICO_DEFAULT_SPI_SCK_PIN, GPIO_FUNC_SPI));
203
204         // Chip select is active-low, so we'll initialise it to a driven-high state
205         gpio_init(PICO_DEFAULT_SPI_CSN_PIN);
206         gpio_set_dir(PICO_DEFAULT_SPI_CSN_PIN, GPIO_OUT);
207         gpio_put(PICO_DEFAULT_SPI_CSN_PIN, 1);
208         // Make the CS pin available to picotool
209         bi_decl(bi_1pin_with_name(PICO_DEFAULT_SPI_CSN_PIN, "SPI CS"));
210
211         // See if SPI is working - interrogate the device for its I2C ID number, should be 0x60
212         uint8_t id;
213         read_registers(0xD0, &id, 1);
214         printf("Chip ID is 0x%x\n", id);
```

```
215
216     read_compensation_parameters();
217
218     write_register(0xF2, 0x1); // Humidity oversampling register - going for x1
219     write_register(0xF4, 0x27);// Set rest of oversampling modes and run mode to normal
220
221     int32_t humidity, pressure, temperature;
222
223     while (1) {
224         bme280_read_raw(&humidity, &pressure, &temperature);
225
226         // These are the raw numbers from the chip, so we need to run through the
227         // compensations to get human understakable numbers
228         pressure = compensate_pressure(pressure);
229         temperature = compensate_temp(temperature);
230         humidity = compensate_humidity(humidity);
231
232         printf("Humidity = %.2f%%\n", humidity / 1024.0);
233         printf("Pressure = %dPa\n", pressure);
234         printf("Temp. = %.2fC\n", temperature / 100.0);
235
236         sleep_ms(1000);
237     }
238
239     return 0;
240 #endif
241 }
```

Bill of Materials

Table 13. A list of materials required for the example

Item	Quantity	Details
Breadboard	1	generic part
Raspberry Pi Pico	1	https://www.raspberrypi.com/products/raspberry-pi-pico/
BME280 board	1	generic part
M/M Jumper wires	6	generic part

Attaching a MPU9250 accelerometer/gyroscope via SPI

This example code shows how to interface the Raspberry Pi Pico to the MPU9250 accelerometer/gyroscope board. The particular device used can be interfaced via I2C or SPI, we are using SPI, and interfacing at 3.3v.

 NOTE

> This is a very basic example, and only recovers raw data from the sensor. There are various calibration options available that should be used to ensure that the final results are accurate. It is also possible to wire up the interrupt pin to a GPIO and read data only when it is ready, rather than using the polling approach in the example.

Wiring information

Wiring up the device requires 6 jumpers as follows:

- GPIO 4 (pin 6) MISO/spi0_rx → ADO on MPU9250 board
- GPIO 5 (pin 7) Chip select → NCS on MPU9250 board
- GPIO 6 (pin 9) SCK/spi0_sclk → SCL on MPU9250 board
- GPIO 7 (pin 10) MOSI/spi0_tx → SDA on MPU9250 board
- 3.3v (pin 36) → VCC on MPU9250 board
- GND (pin 38) → GND on MPU9250 board

The example here uses SPI port 0. Power is supplied from the 3.3V pin.

> **NOTE**
>
> There are many different manufacturers who sell boards with the MPU9250. Whilst they all appear slightly different, they all have, at least, the same 6 pins required to power and communicate. When wiring up a board that is different to the one in the diagram, ensure you connect up as described in the previous paragraph.

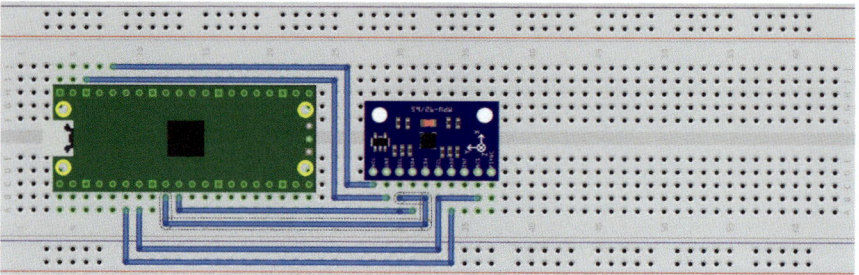

Figure 15. Wiring Diagram for MPU9250.

List of Files

CMakeLists.txt

CMake file to incorporate the example in to the examples build tree.

Pico Examples: *https://github.com/raspberrypi/pico-examples/blob/master/spi/mpu9250_spi/CMakeLists.txt*

```
1  add_executable(mpu9250_spi
2          mpu9250_spi.c
3          )
4
5  # pull in common dependencies and additional spi hardware support
6  target_link_libraries(mpu9250_spi pico_stdlib hardware_spi)
7
8  # create map/bin/hex file etc.
9  pico_add_extra_outputs(mpu9250_spi)
10
11 # add url via pico_set_program_url
12 example_auto_set_url(mpu9250_spi)
```

mpu9250_spi.c

The example code.

Pico Examples: *https://github.com/raspberrypi/pico-examples/blob/master/spi/mpu9250_spi/mpu9250_spi.c*

```
1  /**
2   * Copyright (c) 2020 Raspberry Pi (Trading) Ltd.
3   *
```

```c
 4   * SPDX-License-Identifier: BSD-3-Clause
 5   */
 6
 7  #include <stdio.h>
 8  #include <string.h>
 9  #include "pico/stdlib.h"
10  #include "pico/binary_info.h"
11  #include "hardware/spi.h"
12
13  /* Example code to talk to a MPU9250 MEMS accelerometer and gyroscope.
14     Ignores the magnetometer, that is left as a exercise for the reader.
15
16     This is taking to simple approach of simply reading registers. It's perfectly
17     possible to link up an interrupt line and set things up to read from the
18     inbuilt FIFO to make it more useful.
19
20     NOTE: Ensure the device is capable of being driven at 3.3v NOT 5v. The Pico
21     GPIO (and therefor SPI) cannot be used at 5v.
22
23     You will need to use a level shifter on the I2C lines if you want to run the
24     board at 5v.
25
26     Connections on Raspberry Pi Pico board and a generic MPU9250 board, other
27     boards may vary.
28
29     GPIO 4 (pin 6) MISO/spi0_rx-> ADO on MPU9250 board
30     GPIO 5 (pin 7) Chip select -> NCS on MPU9250 board
31     GPIO 6 (pin 9) SCK/spi0_sclk -> SCL on MPU9250 board
32     GPIO 7 (pin 10) MOSI/spi0_tx -> SDA on MPU9250 board
33     3.3v (pin 36) -> VCC on MPU9250 board
34     GND (pin 38)  -> GND on MPU9250 board
35
36     Note: SPI devices can have a number of different naming schemes for pins. See
37     the Wikipedia page at https://en.wikipedia.org/wiki/Serial_Peripheral_Interface
38     for variations.
39     The particular device used here uses the same pins for I2C and SPI, hence the
40     using of I2C names
41  */
42
43  #define PIN_MISO 4
44  #define PIN_CS   5
45  #define PIN_SCK  6
46  #define PIN_MOSI 7
47
48  #define SPI_PORT spi0
49  #define READ_BIT 0x80
50
51  static inline void cs_select() {
52      asm volatile("nop \n nop \n nop");
53      gpio_put(PIN_CS, 0);  // Active low
54      asm volatile("nop \n nop \n nop");
55  }
56
57  static inline void cs_deselect() {
58      asm volatile("nop \n nop \n nop");
59      gpio_put(PIN_CS, 1);
60      asm volatile("nop \n nop \n nop");
61  }
62
63  static void mpu9250_reset() {
64      // Two byte reset. First byte register, second byte data
65      // There are a load more options to set up the device in different ways that could be added here
```

```c
66        uint8_t buf[] = {0x6B, 0x00};
67        cs_select();
68        spi_write_blocking(SPI_PORT, buf, 2);
69        cs_deselect();
70  }
71
72
73  static void read_registers(uint8_t reg, uint8_t *buf, uint16_t len) {
74      // For this particular device, we send the device the register we want to read
75      // first, then subsequently read from the device. The register is auto incrementing
76      // so we don't need to keep sending the register we want, just the first.
77
78      reg |= READ_BIT;
79      cs_select();
80      spi_write_blocking(SPI_PORT, &reg, 1);
81      sleep_ms(10);
82      spi_read_blocking(SPI_PORT, 0, buf, len);
83      cs_deselect();
84      sleep_ms(10);
85  }
86
87
88  static void mpu9250_read_raw(int16_t accel[3], int16_t gyro[3], int16_t *temp) {
89      uint8_t buffer[6];
90
91      // Start reading acceleration registers from register 0x3B for 6 bytes
92      read_registers(0x3B, buffer, 6);
93
94      for (int i = 0; i < 3; i++) {
95          accel[i] = (buffer[i * 2] << 8 | buffer[(i * 2) + 1]);
96      }
97
98      // Now gyro data from reg 0x43 for 6 bytes
99      read_registers(0x43, buffer, 6);
100
101     for (int i = 0; i < 3; i++) {
102         gyro[i] = (buffer[i * 2] << 8 | buffer[(i * 2) + 1]);;
103     }
104
105     // Now temperature from reg 0x41 for 2 bytes
106     read_registers(0x41, buffer, 2);
107
108     *temp = buffer[0] << 8 | buffer[1];
109 }
110
111 int main() {
112     stdio_init_all();
113
114     printf("Hello, MPU9250! Reading raw data from registers via SPI...\n");
115
116     // This example will use SPI0 at 0.5MHz.
117     spi_init(SPI_PORT, 500 * 1000);
118     gpio_set_function(PIN_MISO, GPIO_FUNC_SPI);
119     gpio_set_function(PIN_SCK, GPIO_FUNC_SPI);
120     gpio_set_function(PIN_MOSI, GPIO_FUNC_SPI);
121     // Make the SPI pins available to picotool
122     bi_decl(bi_3pins_with_func(PIN_MISO, PIN_MOSI, PIN_SCK, GPIO_FUNC_SPI));
123
124     // Chip select is active-low, so we'll initialise it to a driven-high state
125     gpio_init(PIN_CS);
126     gpio_set_dir(PIN_CS, GPIO_OUT);
127     gpio_put(PIN_CS, 1);
128     // Make the CS pin available to picotool
```

```
129        bi_decl(bi_1pin_with_name(PIN_CS, "SPI CS"));
130
131    mpu9250_reset();
132
133    // See if SPI is working - interrogate the device for its I2C ID number, should be 0x71
134    uint8_t id;
135    read_registers(0x75, &id, 1);
136    printf("I2C address is 0x%x\n", id);
137
138    int16_t acceleration[3], gyro[3], temp;
139
140    while (1) {
141        mpu9250_read_raw(acceleration, gyro, &temp);
142
143        // These are the raw numbers from the chip, so will need tweaking to be really useful.
144        // See the datasheet for more information
145        printf("Acc. X = %d, Y = %d, Z = %d\n", acceleration[0], acceleration[1], acceleration[2]);
146        printf("Gyro. X = %d, Y = %d, Z = %d\n", gyro[0], gyro[1], gyro[2]);
147        // Temperature is simple so use the datasheet calculation to get deg C.
148        // Note this is chip temperature.
149        printf("Temp. = %f\n", (temp / 340.0) + 36.53);
150
151        sleep_ms(100);
152    }
153
154    return 0;
155 }
```

Bill of Materials

Table 14. A list of materials required for the example

Item	Quantity	Details
Breadboard	1	generic part
Raspberry Pi Pico	1	https://www.raspberrypi.com/products/raspberry-pi-pico/
MPU9250 board	1	generic part
M/M Jumper wires	6	generic part

Attaching a MPU6050 accelerometer/gyroscope via I2C

This example code shows how to interface the Raspberry Pi Pico to the MPU6050 accelerometer/gyroscope board. This device uses I2C for communications, and most MPU6050 parts are happy running at either 3.3 or 5v. The Raspberry Pi RP2040 GPIO's work at 3.3v so that is what the example uses.

> **NOTE**
>
> This is a very basic example, and only recovers raw data from the sensor. There are various calibration options available that should be used to ensure that the final results are accurate. It is also possible to wire up the interrupt pin to a GPIO and read data only when it is ready, rather than using the polling approach in the example.

Wiring information

Wiring up the device requires 4 jumpers, to connect VCC (3.3v), GND, SDA and SCL. The example here uses I2C port 0, which is assigned to GPIO 4 (SDA) and 5 (SCL) in software. Power is supplied from the 3.3V pin.

> **NOTE**
>
> There are many different manufacturers who sell boards with the MPU6050. Whilst they all appear slightly different, they all have, at least, the same 4 pins required to power and communicate. When wiring up a board that is different to the one in the diagram, ensure you connect up as described in the previous paragraph.

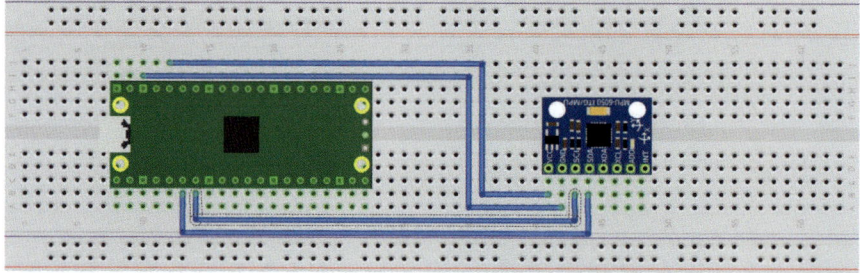

Figure 16. Wiring Diagram for MPU6050.

List of Files

CMakeLists.txt

   CMake file to incorporate the example in to the examples build tree.

Pico Examples: *https://github.com/raspberrypi/pico-examples/blob/master/i2c/mpu6050_i2c/CMakeLists.txt*

```
1  add_executable(mpu6050_i2c
2          mpu6050_i2c.c
3          )
4
5  # pull in common dependencies and additional i2c hardware support
6  target_link_libraries(mpu6050_i2c pico_stdlib hardware_i2c)
7
8  # create map/bin/hex file etc.
9  pico_add_extra_outputs(mpu6050_i2c)
10
11 # add url via pico_set_program_url
12 example_auto_set_url(mpu6050_i2c)
```

mpu6050_i2c.c

   The example code.

Pico Examples: *https://github.com/raspberrypi/pico-examples/blob/master/i2c/mpu6050_i2c/mpu6050_i2c.c*

```
1  /**
```

```c
 2  * Copyright (c) 2020 Raspberry Pi (Trading) Ltd.
 3  *
 4  * SPDX-License-Identifier: BSD-3-Clause
 5  */
 6
 7 #include <stdio.h>
 8 #include <string.h>
 9 #include "pico/stdlib.h"
10 #include "pico/binary_info.h"
11 #include "hardware/i2c.h"
12
13 /* Example code to talk to a MPU6050 MEMS accelerometer and gyroscope
14
15    This is taking to simple approach of simply reading registers. It's perfectly
16    possible to link up an interrupt line and set things up to read from the
17    inbuilt FIFO to make it more useful.
18
19    NOTE: Ensure the device is capable of being driven at 3.3v NOT 5v. The Pico
20    GPIO (and therefor I2C) cannot be used at 5v.
21
22    You will need to use a level shifter on the I2C lines if you want to run the
23    board at 5v.
24
25    Connections on Raspberry Pi Pico board, other boards may vary.
26
27    GPIO PICO_DEFAULT_I2C_SDA_PIN (On Pico this is GP4 (pin 6)) -> SDA on MPU6050 board
28    GPIO PICO_DEFAULT_I2C_SCL_PIN (On Pico this is GP5 (pin 7)) -> SCL on MPU6050 board
29    3.3v (pin 36) -> VCC on MPU6050 board
30    GND (pin 38)  -> GND on MPU6050 board
31 */
32
33 // By default these devices  are on bus address 0x68
34 static int addr = 0x68;
35
36 #ifdef i2c_default
37 static void mpu6050_reset() {
38     // Two byte reset. First byte register, second byte data
39     // There are a load more options to set up the device in different ways that could be added here
40     uint8_t buf[] = {0x6B, 0x00};
41     i2c_write_blocking(i2c_default, addr, buf, 2, false);
42 }
43
44 static void mpu6050_read_raw(int16_t accel[3], int16_t gyro[3], int16_t *temp) {
45     // For this particular device, we send the device the register we want to read
46     // first, then subsequently read from the device. The register is auto incrementing
47     // so we don't need to keep sending the register we want, just the first.
48
49     uint8_t buffer[6];
50
51     // Start reading acceleration registers from register 0x3B for 6 bytes
52     uint8_t val = 0x3B;
53     i2c_write_blocking(i2c_default, addr, &val, 1, true); // true to keep master control of bus
54     i2c_read_blocking(i2c_default, addr, buffer, 6, false);
55
56     for (int i = 0; i < 3; i++) {
57         accel[i] = (buffer[i * 2] << 8 | buffer[(i * 2) + 1]);
58     }
59
60     // Now gyro data from reg 0x43 for 6 bytes
61     // The register is auto incrementing on each read
62     val = 0x43;
```

```c
63        i2c_write_blocking(i2c_default, addr, &val, 1, true);
64        i2c_read_blocking(i2c_default, addr, buffer, 6, false);  // False - finished with bus
65
66        for (int i = 0; i < 3; i++) {
67            gyro[i] = (buffer[i * 2] << 8 | buffer[(i * 2) + 1]);;
68        }
69
70        // Now temperature from reg 0x41 for 2 bytes
71        // The register is auto incrementing on each read
72        val = 0x41;
73        i2c_write_blocking(i2c_default, addr, &val, 1, true);
74        i2c_read_blocking(i2c_default, addr, buffer, 2, false);  // False - finished with bus
75
76        *temp = buffer[0] << 8 | buffer[1];
77 }
78 #endif
79
80 int main() {
81     stdio_init_all();
82 #if !defined(i2c_default) || !defined(PICO_DEFAULT_I2C_SDA_PIN) ||
   !defined(PICO_DEFAULT_I2C_SCL_PIN)
83     #warning i2c/mpu6050_i2c example requires a board with I2C pins
84     puts("Default I2C pins were not defined");
85 #else
86     printf("Hello, MPU6050! Reading raw data from registers...\n");
87
88     // This example will use I2C0 on the default SDA and SCL pins (4, 5 on a Pico)
89     i2c_init(i2c_default, 400 * 1000);
90     gpio_set_function(PICO_DEFAULT_I2C_SDA_PIN, GPIO_FUNC_I2C);
91     gpio_set_function(PICO_DEFAULT_I2C_SCL_PIN, GPIO_FUNC_I2C);
92     gpio_pull_up(PICO_DEFAULT_I2C_SDA_PIN);
93     gpio_pull_up(PICO_DEFAULT_I2C_SCL_PIN);
94     // Make the I2C pins available to picotool
95     bi_decl(bi_2pins_with_func(PICO_DEFAULT_I2C_SDA_PIN, PICO_DEFAULT_I2C_SCL_PIN,
   GPIO_FUNC_I2C));
96
97     mpu6050_reset();
98
99     int16_t acceleration[3], gyro[3], temp;
100
101     while (1) {
102         mpu6050_read_raw(acceleration, gyro, &temp);
103
104         // These are the raw numbers from the chip, so will need tweaking to be really
   useful.
105         // See the datasheet for more information
106         printf("Acc. X = %d, Y = %d, Z = %d\n", acceleration[0], acceleration[1],
   acceleration[2]);
107         printf("Gyro. X = %d, Y = %d, Z = %d\n", gyro[0], gyro[1], gyro[2]);
108         // Temperature is simple so use the datasheet calculation to get deg C.
109         // Note this is chip temperature.
110         printf("Temp. = %f\n", (temp / 340.0) + 36.53);
111
112         sleep_ms(100);
113     }
114
115 #endif
116     return 0;
117 }
```

Bill of Materials

Table 15. A list of materials required for the example

Item	Quantity	Details
Breadboard	1	generic part
Raspberry Pi Pico	1	https://www.raspberrypi.com/products/raspberry-pi-pico/
MPU6050 board	1	generic part
M/M Jumper wires	4	generic part

Attaching a 16x2 LCD via I2C

This example code shows how to interface the Raspberry Pi Pico to one of the very common 16x2 LCD character displays. The display will need a 3.3V I2C adapter board as this example uses I2C for communications.

> **NOTE**
>
> These LCD displays can also be driven directly using GPIO without the use of an adapter board. That is beyond the scope of this example.

Wiring information

Wiring up the device requires 4 jumpers, to connect VCC (3.3v), GND, SDA and SCL. The example here uses I2C port 0, which is assigned to GPIO 4 (SDA) and 5 (SCL) in software. Power is supplied from the 3.3V pin.

> **WARNING**
>
> Many displays of this type are 5v. If you wish to use a 5v display you will need to use level shifters on the SDA and SCL lines to convert from the 3.3V used by the RP2040. Whilst a 5v display will just about work at 3.3v, the display will be dim.

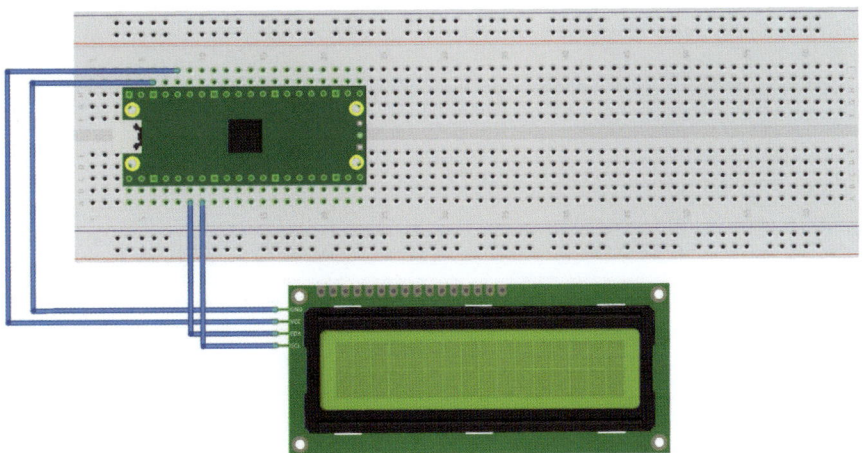

Figure 17. Wiring Diagram for LCD1602A LCD with I2C bridge.

List of Files

CMakeLists.txt

CMake file to incorporate the example in to the examples build tree.

Pico Examples: *https://github.com/raspberrypi/pico-examples/blob/master/i2c/lcd_1602_i2c/CMakeLists.txt*

```
1  add_executable(lcd_1602_i2c
2          lcd_1602_i2c.c
3          )
4
5  # pull in common dependencies and additional i2c hardware support
6  target_link_libraries(lcd_1602_i2c pico_stdlib hardware_i2c)
7
8  # create map/bin/hex file etc.
9  pico_add_extra_outputs(lcd_1602_i2c)
10
11 # add url via pico_set_program_url
12 example_auto_set_url(lcd_1602_i2c)
```

lcd_1602_i2c.c

The example code.

Pico Examples: *https://github.com/raspberrypi/pico-examples/blob/master/i2c/lcd_1602_i2c/lcd_1602_i2c.c*

```
1  /**
2   * Copyright (c) 2020 Raspberry Pi (Trading) Ltd.
3   *
4   * SPDX-License-Identifier: BSD-3-Clause
5   */
6
7  #include <stdio.h>
8  #include <string.h>
9  #include "pico/stdlib.h"
10 #include "hardware/i2c.h"
11 #include "pico/binary_info.h"
12
13 /* Example code to drive a 16x2 LCD panel via a I2C bridge chip (e.g. PCF8574)
14
15    NOTE: The panel must be capable of being driven at 3.3v NOT 5v. The Pico
16    GPIO (and therefor I2C) cannot be used at 5v.
17
18    You will need to use a level shifter on the I2C lines if you want to run the
19    board at 5v.
20
21    Connections on Raspberry Pi Pico board, other boards may vary.
22
23    GPIO 4 (pin 6)-> SDA on LCD bridge board
24    GPIO 5 (pin 7)-> SCL on LCD bridge board
25    3.3v (pin 36) -> VCC on LCD bridge board
26    GND (pin 38)  -> GND on LCD bridge board
27 */
28 // commands
29 const int LCD_CLEARDISPLAY = 0x01;
30 const int LCD_RETURNHOME = 0x02;
31 const int LCD_ENTRYMODESET = 0x04;
32 const int LCD_DISPLAYCONTROL = 0x08;
33 const int LCD_CURSORSHIFT = 0x10;
34 const int LCD_FUNCTIONSET = 0x20;
35 const int LCD_SETCGRAMADDR = 0x40;
36 const int LCD_SETDDRAMADDR = 0x80;
37
```

```c
38  // flags for display entry mode
39  const int LCD_ENTRYSHIFTINCREMENT = 0x01;
40  const int LCD_ENTRYLEFT = 0x02;
41
42  // flags for display and cursor control
43  const int LCD_BLINKON = 0x01;
44  const int LCD_CURSORON = 0x02;
45  const int LCD_DISPLAYON = 0x04;
46
47  // flags for display and cursor shift
48  const int LCD_MOVERIGHT = 0x04;
49  const int LCD_DISPLAYMOVE = 0x08;
50
51  // flags for function set
52  const int LCD_5x10DOTS = 0x04;
53  const int LCD_2LINE = 0x08;
54  const int LCD_8BITMODE = 0x10;
55
56  // flag for backlight control
57  const int LCD_BACKLIGHT = 0x08;
58
59  const int LCD_ENABLE_BIT = 0x04;
60
61  // By default these LCD display drivers are on bus address 0x27
62  static int addr = 0x27;
63
64  // Modes for lcd_send_byte
65  #define LCD_CHARACTER  1
66  #define LCD_COMMAND    0
67
68  #define MAX_LINES      2
69  #define MAX_CHARS      16
70
71  /* Quick helper function for single byte transfers */
72  void i2c_write_byte(uint8_t val) {
73  #ifdef i2c_default
74      i2c_write_blocking(i2c_default, addr, &val, 1, false);
75  #endif
76  }
77
78  void lcd_toggle_enable(uint8_t val) {
79      // Toggle enable pin on LCD display
80      // We cannot do this too quickly or things don't work
81  #define DELAY_US 600
82      sleep_us(DELAY_US);
83      i2c_write_byte(val | LCD_ENABLE_BIT);
84      sleep_us(DELAY_US);
85      i2c_write_byte(val & ~LCD_ENABLE_BIT);
86      sleep_us(DELAY_US);
87  }
88
89  // The display is sent a byte as two separate nibble transfers
90  void lcd_send_byte(uint8_t val, int mode) {
91      uint8_t high = mode | (val & 0xF0) | LCD_BACKLIGHT;
92      uint8_t low = mode | ((val << 4) & 0xF0) | LCD_BACKLIGHT;
93
94      i2c_write_byte(high);
95      lcd_toggle_enable(high);
96      i2c_write_byte(low);
97      lcd_toggle_enable(low);
98  }
99
100 void lcd_clear(void) {
```

```c
        lcd_send_byte(LCD_CLEARDISPLAY, LCD_COMMAND);
}

// go to location on LCD
void lcd_set_cursor(int line, int position) {
    int val = (line == 0) ? 0x80 + position : 0xC0 + position;
    lcd_send_byte(val, LCD_COMMAND);
}

static void inline lcd_char(char val) {
    lcd_send_byte(val, LCD_CHARACTER);
}

void lcd_string(const char *s) {
    while (*s) {
        lcd_char(*s++);
    }
}

void lcd_init() {
    lcd_send_byte(0x03, LCD_COMMAND);
    lcd_send_byte(0x03, LCD_COMMAND);
    lcd_send_byte(0x03, LCD_COMMAND);
    lcd_send_byte(0x02, LCD_COMMAND);

    lcd_send_byte(LCD_ENTRYMODESET | LCD_ENTRYLEFT, LCD_COMMAND);
    lcd_send_byte(LCD_FUNCTIONSET | LCD_2LINE, LCD_COMMAND);
    lcd_send_byte(LCD_DISPLAYCONTROL | LCD_DISPLAYON, LCD_COMMAND);
    lcd_clear();
}

int main() {
#if !defined(i2c_default) || !defined(PICO_DEFAULT_I2C_SDA_PIN) || !defined(PICO_DEFAULT_I2C_SCL_PIN)
    #warning i2c/lcd_1602_i2c example requires a board with I2C pins
#else
    // This example will use I2C0 on the default SDA and SCL pins (4, 5 on a Pico)
    i2c_init(i2c_default, 100 * 1000);
    gpio_set_function(PICO_DEFAULT_I2C_SDA_PIN, GPIO_FUNC_I2C);
    gpio_set_function(PICO_DEFAULT_I2C_SCL_PIN, GPIO_FUNC_I2C);
    gpio_pull_up(PICO_DEFAULT_I2C_SDA_PIN);
    gpio_pull_up(PICO_DEFAULT_I2C_SCL_PIN);
    // Make the I2C pins available to picotool
    bi_decl(bi_2pins_with_func(PICO_DEFAULT_I2C_SDA_PIN, PICO_DEFAULT_I2C_SCL_PIN, GPIO_FUNC_I2C));

    lcd_init();

    static char *message[] =
            {
                    "RP2040 by", "Raspberry Pi",
                    "A brand new", "microcontroller",
                    "Twin core M0", "Full C SDK",
                    "More power in", "your product",
                    "More beans", "than Heinz!"
            };

    while (1) {
        for (int m = 0; m < sizeof(message) / sizeof(message[0]); m += MAX_LINES) {
            for (int line = 0; line < MAX_LINES; line++) {
                lcd_set_cursor(line, (MAX_CHARS / 2) - strlen(message[m + line]) / 2);
                lcd_string(message[m + line]);
            }
```

```
162                sleep_ms(2000);
163                lcd_clear();
164            }
165        }
166
167        return 0;
168 #endif
169 }
```

Bill of Materials

Table 16. A list of materials required for the example

Item	Quantity	Details
Breadboard	1	generic part
Raspberry Pi Pico	1	https://www.raspberrypi.com/products/raspberry-pi-pico/
1602A based LCD panel 3.3v	1	generic part
1602A to I2C bridge device 3.3v	1	generic part
M/M Jumper wires	4	generic part

Attaching a BMP280 temp/pressure sensor via I2C

This example code shows how to interface the Raspberry Pi Pico with the popular BMP280 temperature and air pressure sensor manufactured by Bosch. A similar variant, the BME280, exists that can also measure humidity. There is another example that uses the BME280 device but talks to it via SPI as opposed to I2C.

The code reads data from the sensor's registers every 500 milliseconds and prints it via the onboard UART. This example operates the BMP280 in *normal* mode, meaning that the device continuously cycles between a measurement period and a standby period at a regular interval we can set. This has the advantage that subsequent reads do not require configuration register writes and is the recommended mode of operation to filter out short-term disturbances.

 TIP

> The BMP280 is highly configurable with 3 modes of operation, various oversampling levels, and 5 filter settings. Find the datasheet online (https://www.bosch-sensortec.com/media/boschsensortec/downloads/datasheets/bst-bmp280-ds001.pdf) to explore all of its capabilities beyond the simple example given here.

Wiring information

Wiring up the device requires 4 jumpers, to connect VCC (3.3v), GND, SDA and SCL. The example here uses the default I2C port 0, which is assigned to GPIO 4 (SDA) and 5 (SCL) in software. Power is supplied from the 3.3V pin from the Pico.

> **WARNING**
>
> The BMP280 has a maximum supply voltage rating of 3.6V. Most breakout boards have voltage regulators that will allow a range of input voltages of 2-6V, but make sure to check beforehand.

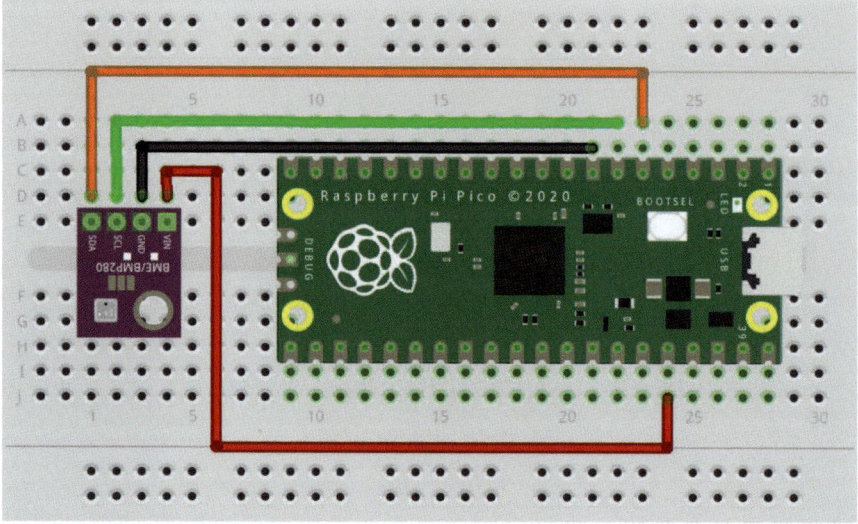

Figure 18. Wiring Diagram for BMP280 sensor via I2C.

List of Files

CMakeLists.txt

CMake file to incorporate the example into the examples build tree.

Pico Examples: https://github.com/raspberrypi/pico-examples/blob/master/i2c/bmp280_i2c/CMakeLists.txt

```
 1 add_executable(bmp280_i2c
 2         bmp280_i2c.c
 3         )
 4
 5 # pull in common dependencies and additional i2c hardware support
 6 target_link_libraries(bmp280_i2c pico_stdlib hardware_i2c)
 7
 8 # create map/bin/hex file etc.
 9 pico_add_extra_outputs(bmp280_i2c)
10
11 # add url via pico_set_program_url
12 example_auto_set_url(bmp280_i2c)
```

bmp280_i2c.c

The example code.

Pico Examples: https://github.com/raspberrypi/pico-examples/blob/master/i2c/bmp280_i2c/bmp280_i2c.c

```
1 /**
2  * Copyright (c) 2021 Raspberry Pi (Trading) Ltd.
3  *
4  * SPDX-License-Identifier: BSD-3-Clause
5  **/
6
```

```c
#include <stdio.h>

#include "hardware/i2c.h"
#include "pico/binary_info.h"
#include "pico/stdlib.h"

/* Example code to talk to a BMP280 temperature and pressure sensor

   NOTE: Ensure the device is capable of being driven at 3.3v NOT 5v. The Pico
   GPIO (and therefore I2C) cannot be used at 5v.

   You will need to use a level shifter on the I2C lines if you want to run the
   board at 5v.

   Connections on Raspberry Pi Pico board, other boards may vary.

   GPIO PICO_DEFAULT_I2C_SDA_PIN (on Pico this is GP4 (pin 6)) -> SDA on BMP280
   board
   GPIO PICO_DEFAULT_I2C_SCK_PIN (on Pico this is GP5 (pin 7)) -> SCL on
   BMP280 board
   3.3v (pin 36) -> VCC on BMP280 board
   GND (pin 38)  -> GND on BMP280 board
*/

// device has default bus address of 0x76
#define ADDR _u(0x76)

// hardware registers
#define REG_CONFIG _u(0xF5)
#define REG_CTRL_MEAS _u(0xF4)
#define REG_RESET _u(0xE0)

#define REG_TEMP_XLSB _u(0xFC)
#define REG_TEMP_LSB _u(0xFB)
#define REG_TEMP_MSB _u(0xFA)

#define REG_PRESSURE_XLSB _u(0xF9)
#define REG_PRESSURE_LSB _u(0xF8)
#define REG_PRESSURE_MSB _u(0xF7)

// calibration registers
#define REG_DIG_T1_LSB _u(0x88)
#define REG_DIG_T1_MSB _u(0x89)
#define REG_DIG_T2_LSB _u(0x8A)
#define REG_DIG_T2_MSB _u(0x8B)
#define REG_DIG_T3_LSB _u(0x8C)
#define REG_DIG_T3_MSB _u(0x8D)
#define REG_DIG_P1_LSB _u(0x8E)
#define REG_DIG_P1_MSB _u(0x8F)
#define REG_DIG_P2_LSB _u(0x90)
#define REG_DIG_P2_MSB _u(0x91)
#define REG_DIG_P3_LSB _u(0x92)
#define REG_DIG_P3_MSB _u(0x93)
#define REG_DIG_P4_LSB _u(0x94)
#define REG_DIG_P4_MSB _u(0x95)
#define REG_DIG_P5_LSB _u(0x96)
#define REG_DIG_P5_MSB _u(0x97)
#define REG_DIG_P6_LSB _u(0x98)
#define REG_DIG_P6_MSB _u(0x99)
#define REG_DIG_P7_LSB _u(0x9A)
#define REG_DIG_P7_MSB _u(0x9B)
#define REG_DIG_P8_LSB _u(0x9C)
#define REG_DIG_P8_MSB _u(0x9D)
```

```c
70  #define REG_DIG_P9_LSB _u(0x9E)
71  #define REG_DIG_P9_MSB _u(0x9F)
72
73  // number of calibration registers to be read
74  #define NUM_CALIB_PARAMS 24
75
76  struct bmp280_calib_param {
77      // temperature params
78      uint16_t dig_t1;
79      int16_t dig_t2;
80      int16_t dig_t3;
81
82      // pressure params
83      uint16_t dig_p1;
84      int16_t dig_p2;
85      int16_t dig_p3;
86      int16_t dig_p4;
87      int16_t dig_p5;
88      int16_t dig_p6;
89      int16_t dig_p7;
90      int16_t dig_p8;
91      int16_t dig_p9;
92  };
93
94  #ifdef i2c_default
95  void bmp280_init() {
96      // use the "handheld device dynamic" optimal setting (see datasheet)
97      uint8_t buf[2];
98
99      // 500ms sampling time, x16 filter
100     const uint8_t reg_config_val = ((0x04 << 5) | (0x05 << 2)) & 0xFC;
101
102     // send register number followed by its corresponding value
103     buf[0] = REG_CONFIG;
104     buf[1] = reg_config_val;
105     i2c_write_blocking(i2c_default, ADDR, buf, 2, false);
106
107     // osrs_t x1, osrs_p x4, normal mode operation
108     const uint8_t reg_ctrl_meas_val = (0x01 << 5) | (0x03 << 2) | (0x03);
109     buf[0] = REG_CTRL_MEAS;
110     buf[1] = reg_ctrl_meas_val;
111     i2c_write_blocking(i2c_default, ADDR, buf, 2, false);
112 }
113
114 void bmp280_read_raw(int32_t* temp, int32_t* pressure) {
115     // BMP280 data registers are auto-incrementing and we have 3 temperature and
116     // pressure registers each, so we start at 0xF7 and read 6 bytes to 0xFC
117     // note: normal mode does not require further ctrl_meas and config register writes
118
119     uint8_t buf[6];
120     uint8_t reg = REG_PRESSURE_MSB;
121     i2c_write_blocking(i2c_default, ADDR, &reg, 1, true);  // true to keep master control of bus
122     i2c_read_blocking(i2c_default, ADDR, buf, 6, false);   // false - finished with bus
123
124     // store the 20 bit read in a 32 bit signed integer for conversion
125     *pressure = (buf[0] << 12) | (buf[1] << 4) | (buf[2] >> 4);
126     *temp = (buf[3] << 12) | (buf[4] << 4) | (buf[5] >> 4);
127 }
128
129 void bmp280_reset() {
130     // reset the device with the power-on-reset procedure
131     uint8_t buf[2] = { REG_RESET, 0xB6 };
```

```
132        i2c_write_blocking(i2c_default, ADDR, buf, 2, false);
133 }
134
135 // intermediate function that calculates the fine resolution temperature
136 // used for both pressure and temperature conversions
137 int32_t bmp280_convert(int32_t temp, struct bmp280_calib_param* params) {
138     // use the 32-bit fixed point compensation implementation given in the
139     // datasheet
140
141     int32_t var1, var2;
142     var1 = ((((temp >> 3) - ((int32_t)params->dig_t1 << 1))) * ((int32_t)params->dig_t2)) >> 11;
143     var2 = (((((temp >> 4) - ((int32_t)params->dig_t1)) * ((temp >> 4) - ((int32_t)params->dig_t1))) >> 12) * ((int32_t)params->dig_t3)) >> 14;
144     return var1 + var2;
145 }
146
147 int32_t bmp280_convert_temp(int32_t temp, struct bmp280_calib_param* params) {
148     // uses the BMP280 calibration parameters to compensate the temperature value read from its registers
149     int32_t t_fine = bmp280_convert(temp, params);
150     return (t_fine * 5 + 128) >> 8;
151 }
152
153 int32_t bmp280_convert_pressure(int32_t pressure, int32_t temp, struct bmp280_calib_param* params) {
154     // uses the BMP280 calibration parameters to compensate the pressure value read from its registers
155
156     int32_t t_fine = bmp280_convert(temp, params);
157
158     int32_t var1, var2;
159     uint32_t converted = 0.0;
160     var1 = (((int32_t)t_fine) >> 1) - (int32_t)64000;
161     var2 = (((var1 >> 2) * (var1 >> 2)) >> 11) * ((int32_t)params->dig_p6);
162     var2 += ((var1 * ((int32_t)params->dig_p5)) << 1);
163     var2 = (var2 >> 2) + (((int32_t)params->dig_p4) << 16);
164     var1 = (((params->dig_p3 * (((var1 >> 2) * (var1 >> 2)) >> 13)) >> 3) + ((((int32_t)params->dig_p2) * var1) >> 1)) >> 18;
165     var1 = ((((32768 + var1)) * ((int32_t)params->dig_p1)) >> 15);
166     if (var1 == 0) {
167         return 0;  // avoid exception caused by division by zero
168     }
169     converted = (((uint32_t)(((int32_t)1048576) - pressure) - (var2 >> 12))) * 3125;
170     if (converted < 0x80000000) {
171         converted = (converted << 1) / ((uint32_t)var1);
172     } else {
173         converted = (converted / (uint32_t)var1) * 2;
174     }
175     var1 = (((int32_t)params->dig_p9) * ((int32_t)(((converted >> 3) * (converted >> 3)) >> 13))) >> 12;
176     var2 = (((int32_t)(converted >> 2)) * ((int32_t)params->dig_p8)) >> 13;
177     converted = (uint32_t)((int32_t)converted + ((var1 + var2 + params->dig_p7) >> 4));
178     return converted;
179 }
180
181 void bmp280_get_calib_params(struct bmp280_calib_param* params) {
182     // raw temp and pressure values need to be calibrated according to
183     // parameters generated during the manufacturing of the sensor
184     // there are 3 temperature params, and 9 pressure params, each with a LSB
185     // and MSB register, so we read from 24 registers
186
187     uint8_t buf[NUM_CALIB_PARAMS] = { 0 };
```

```c
188        uint8_t reg = REG_DIG_T1_LSB;
189        i2c_write_blocking(i2c_default, ADDR, &reg, 1, true);  // true to keep master control of bus
190        // read in one go as register addresses auto-increment
191        i2c_read_blocking(i2c_default, ADDR, buf, NUM_CALIB_PARAMS, false);  // false, we're done reading
192
193        // store these in a struct for later use
194        params->dig_t1 = (uint16_t)(buf[1] << 8) | buf[0];
195        params->dig_t2 = (int16_t)(buf[3] << 8) | buf[2];
196        params->dig_t3 = (int16_t)(buf[5] << 8) | buf[4];
197
198        params->dig_p1 = (uint16_t)(buf[7] << 8) | buf[6];
199        params->dig_p2 = (int16_t)(buf[9] << 8) | buf[8];
200        params->dig_p3 = (int16_t)(buf[11] << 8) | buf[10];
201        params->dig_p4 = (int16_t)(buf[13] << 8) | buf[12];
202        params->dig_p5 = (int16_t)(buf[15] << 8) | buf[14];
203        params->dig_p6 = (int16_t)(buf[17] << 8) | buf[16];
204        params->dig_p7 = (int16_t)(buf[19] << 8) | buf[18];
205        params->dig_p8 = (int16_t)(buf[21] << 8) | buf[20];
206        params->dig_p9 = (int16_t)(buf[23] << 8) | buf[22];
207    }
208
209    #endif
210
211    int main() {
212        stdio_init_all();
213
214    #if !defined(i2c_default) || !defined(PICO_DEFAULT_I2C_SDA_PIN) || !defined(PICO_DEFAULT_I2C_SCL_PIN)
215        #warning i2c / bmp280_i2c example requires a board with I2C pins
216            puts("Default I2C pins were not defined");
217    #else
218        // useful information for picotool
219        bi_decl(bi_2pins_with_func(PICO_DEFAULT_I2C_SDA_PIN, PICO_DEFAULT_I2C_SCL_PIN, GPIO_FUNC_I2C));
220        bi_decl(bi_program_description("BMP280 I2C example for the Raspberry Pi Pico"));
221
222        printf("Hello, BMP280! Reading temperaure and pressure values from sensor...\n");
223
224        // I2C is "open drain", pull ups to keep signal high when no data is being sent
225        i2c_init(i2c_default, 100 * 1000);
226        gpio_set_function(PICO_DEFAULT_I2C_SDA_PIN, GPIO_FUNC_I2C);
227        gpio_set_function(PICO_DEFAULT_I2C_SCL_PIN, GPIO_FUNC_I2C);
228        gpio_pull_up(PICO_DEFAULT_I2C_SDA_PIN);
229        gpio_pull_up(PICO_DEFAULT_I2C_SCL_PIN);
230
231        // configure BMP280
232        bmp280_init();
233
234        // retrieve fixed compensation params
235        struct bmp280_calib_param params;
236        bmp280_get_calib_params(&params);
237
238        int32_t raw_temperature;
239        int32_t raw_pressure;
240
241        sleep_ms(250); // sleep so that data polling and register update don't collide
242        while (1) {
243            bmp280_read_raw(&raw_temperature, &raw_pressure);
244            int32_t temperature = bmp280_convert_temp(raw_temperature, &params);
245            int32_t pressure = bmp280_convert_pressure(raw_pressure, raw_temperature, &params);
246            printf("Pressure = %.3f kPa\n", pressure / 1000.f);
```

```
247            printf("Temp. = %.2f C\n", temperature / 100.f);
248            // poll every 500ms
249            sleep_ms(500);
250        }
251
252    #endif
253        return 0;
254    }
```

Bill of Materials

Table 17. A list of materials required for the example

Item	Quantity	Details
Breadboard	1	generic part
Raspberry Pi Pico	1	https://www.raspberrypi.com/products/raspberry-pi-pico/
BMP280-based breakout board	1	from Pimoroni
M/M Jumper wires	4	generic part

Attaching a LIS3DH Nano Accelerometer via i2c.

This example shows you how to interface the Raspberry Pi Pico to the LIS3DH accelerometer and temperature sensor.

The code reads and displays the acceleration values of the board in the 3 axes and the ambient temperature value. The datasheet for the sensor can be found at https://www.st.com/resource/en/datasheet/cd00274221.pdf. The device is being operated on 'normal mode' and at a frequency of 1.344 kHz (this can be changed by editing the ODR bits of CTRL_REG4). The range of the data is controlled by the FS bit in CTRL_REG4 and is equal to ±2g in this example. The sensitivity depends on the operating mode and data range; exact values can be found on page 10 of the datasheet. In this case, the sensitivity value is 4mg (where g is the value of gravitational acceleration on the surface of Earth). In order to use the auxiliary ADC to read temperature, the we must set the BDU bit to 1 in CTRL_REG4 and the ADC_EN bit to 1 in TEMP_CFG_REG. Temperature is communicated through ADC 3.

 NOTE

> The sensor doesn't have features to eliminate offsets in the data and these will need to be taken into account in the code.

Wiring information

Wiring up the device requires 4 jumpers, to connect VIN, GND, SDA and SCL. The example here uses I2C port 0, which is assigned to GPIO 4 (SDA) and 5 (SCL) in software. Power is supplied from the 3V pin.

Figure 19. Wiring Diagram for LIS3DH.

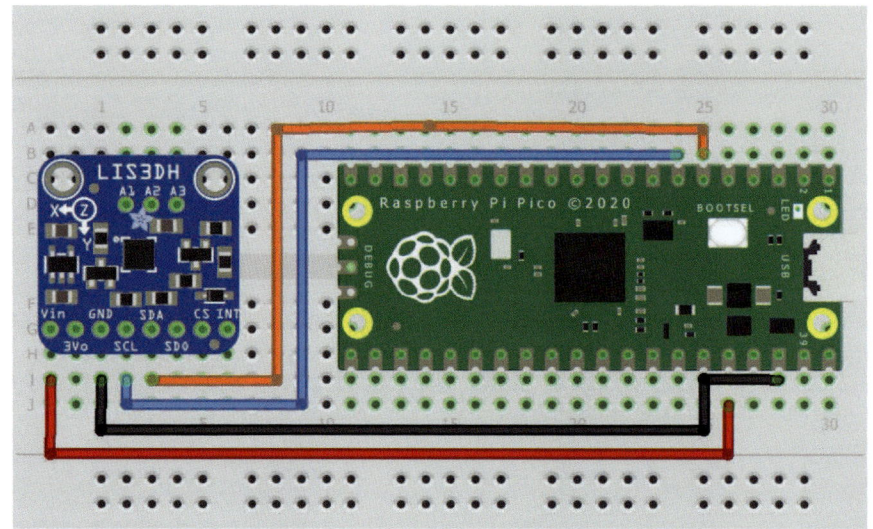

List of Files

CMakeLists.txt

CMake file to incorporate the example in to the examples build tree.

Pico Examples: *https://github.com/raspberrypi/pico-examples/blob/master/i2c/lis3dh_i2c/CMakeLists.txt*

```
1  add_executable(lis3dh_i2c
2          lis3dh_i2c.c
3          )
4
5  # pull in common dependencies and additional i2c hardware support
6  target_link_libraries(lis3dh_i2c pico_stdlib hardware_i2c)
7
8  # create map/bin/hex file etc.
9  pico_add_extra_outputs(lis3dh_i2c)
10
11 # add url via pico_set_program_url
12 example_auto_set_url(lis3dh_i2c)
```

lis3dh_i2c.c

The example code.

Pico Examples: *https://github.com/raspberrypi/pico-examples/blob/master/i2c/lis3dh_i2c/lis3dh_i2c.c*

```
1  /**
2   * Copyright (c) 2020 Raspberry Pi (Trading) Ltd.
3   *
4   * SPDX-License-Identifier: BSD-3-Clause
5   */
6
7  #include <stdio.h>
8  #include <string.h>
9  #include "pico/stdlib.h"
10 #include "pico/binary_info.h"
11 #include "hardware/i2c.h"
12
13 /* Example code to talk to a LIS3DH Mini GPS module.
```

```c
14
15    This example reads data from all 3 axes of the accelerometer and uses an auxillary ADC to
      output temperature values.
16
17    Connections on Raspberry Pi Pico board, other boards may vary.
18
19    GPIO PICO_DEFAULT_I2C_SDA_PIN (On Pico this is 4 (physical pin 6)) -> SDA on LIS3DH board
20    GPIO PICO_DEFAULT_I2C_SCK_PIN (On Pico this is 5 (physical pin 7)) -> SCL on LIS3DH board
21    3.3v (physical pin 36) -> VIN on LIS3DH board
22    GND (physical pin 38)  -> GND on LIS3DH board
23 */
24
25 // By default this device is on bus address 0x18
26
27 const int ADDRESS = 0x18;
28 const uint8_t CTRL_REG_1 = 0x20;
29 const uint8_t CTRL_REG_4 = 0x23;
30 const uint8_t TEMP_CFG_REG = 0xC0;
31
32 #ifdef i2c_default
33
34 void lis3dh_init() {
35     uint8_t buf[2];
36
37     // Turn normal mode and 1.344kHz data rate on
38     buf[0] = CTRL_REG_1;
39     buf[1] = 0x97;
40     i2c_write_blocking(i2c_default, ADDRESS, buf, 2, false);
41
42     // Turn block data update on (for temperature sensing)
43     buf[0] = CTRL_REG_4;
44     buf[1] = 0x80;
45     i2c_write_blocking(i2c_default, ADDRESS, buf, 2, false);
46
47     // Turn auxillary ADC on
48     buf[0] = TEMP_CFG_REG;
49     buf[1] = 0xC0;
50     i2c_write_blocking(i2c_default, ADDRESS, buf, 2, false);
51 }
52
53 void lis3dh_calc_value(uint16_t raw_value, float *final_value, bool isAccel) {
54     // Convert with respect to the value being temperature or acceleration reading
55     float scaling;
56     float senstivity = 0.004f; // g per unit
57
58     if (isAccel == true) {
59         scaling = 64 / senstivity;
60     } else {
61         scaling = 64;
62     }
63
64     // raw_value is signed
65     *final_value = (float) ((int16_t) raw_value) / scaling;
66 }
67
68 void lis3dh_read_data(uint8_t reg, float *final_value, bool IsAccel) {
69     // Read two bytes of data and store in a 16 bit data structure
70     uint8_t lsb;
71     uint8_t msb;
72     uint16_t raw_accel;
73     i2c_write_blocking(i2c_default, ADDRESS, &reg, 1, true);
74     i2c_read_blocking(i2c_default, ADDRESS, &lsb, 1, false);
75
```

```c
 76        reg |= 0x01;
 77        i2c_write_blocking(i2c_default, ADDRESS, &reg, 1, true);
 78        i2c_read_blocking(i2c_default, ADDRESS, &msb, 1, false);
 79
 80        raw_accel = (msb << 8) | lsb;
 81
 82        lis3dh_calc_value(raw_accel, final_value, IsAccel);
 83  }
 84
 85  #endif
 86
 87  int main() {
 88        stdio_init_all();
 89  #if !defined(i2c_default) || !defined(PICO_DEFAULT_I2C_SDA_PIN) ||
     !defined(PICO_DEFAULT_I2C_SCL_PIN)
 90  #warning i2c/lis3dh_i2c example requires a board with I2C pins
 91        puts("Default I2C pins were not defined");
 92  #else
 93        printf("Hello, LIS3DH! Reading raw data from registers...\n");
 94
 95        // This example will use I2C0 on the default SDA and SCL pins (4, 5 on a Pico)
 96        i2c_init(i2c_default, 400 * 1000);
 97        gpio_set_function(PICO_DEFAULT_I2C_SDA_PIN, GPIO_FUNC_I2C);
 98        gpio_set_function(PICO_DEFAULT_I2C_SCL_PIN, GPIO_FUNC_I2C);
 99        gpio_pull_up(PICO_DEFAULT_I2C_SDA_PIN);
100        gpio_pull_up(PICO_DEFAULT_I2C_SCL_PIN);
101        // Make the I2C pins available to picotool
102        bi_decl(bi_2pins_with_func(PICO_DEFAULT_I2C_SDA_PIN, PICO_DEFAULT_I2C_SCL_PIN,
     GPIO_FUNC_I2C));
103
104        float x_accel, y_accel, z_accel, temp;
105
106        lis3dh_init();
107
108        while (1) {
109            lis3dh_read_data(0x28, &x_accel, true);
110            lis3dh_read_data(0x2A, &y_accel, true);
111            lis3dh_read_data(0x2C, &z_accel, true);
112            lis3dh_read_data(0x0C, &temp, false);
113
114            // Display data
115            printf("TEMPERATURE: %.3f%cC\n", temp, 176);
116            // Acceleration is read as a multiple of g (gravitational acceleration on the Earth's
     surface)
117            printf("ACCELERATION VALUES: \n");
118            printf("X acceleration: %.3fg\n", x_accel);
119            printf("Y acceleration: %.3fg\n", y_accel);
120            printf("Z acceleration: %.3fg\n", z_accel);
121
122            sleep_ms(500);
123
124            // Clear terminal
125            printf("\e[1;1H\e[2J");
126        }
127  #endif
128        return 0;
129  }
```

Bill of Materials

Table 18. A list of materials required for the example

Item	Quantity	Details
Breadboard	1	generic part
Raspberry Pi Pico	1	https://www.raspberrypi.com/products/raspberry-pi-pico/
LIS3DH board	1	https://www.adafruit.com/product/2809
M/M Jumper wires	4	generic part

Attaching a MCP9808 digital temperature sensor via I2C

This example code shows how to interface the Raspberry Pi Pico to the MCP9808 digital temperature sensor board.

This example reads the ambient temperature value each second from the sensor and sets upper, lower and critical limits for the temperature and checks if alerts need to be raised. The CONFIG register can also be used to check for an alert if the critical temperature is surpassed.

Wiring information

Wiring up the device requires 4 jumpers, to connect VDD, GND, SDA and SCL. The example here uses I2C port 0, which is assigned to GPIO 4 (SDA) and 5 (SCL) in software. Power is supplied from the VSYS pin.

Figure 20. Wiring Diagram for MCP9808.

List of Files

CMakeLists.txt

CMake file to incorporate the example in to the examples build tree.

Pico Examples: https://github.com/raspberrypi/pico-examples/blob/master/i2c/mcp9808_i2c/CMakeLists.txt

```
1  add_executable(mcp9808_i2c
2          mcp9808_i2c.c
```

```
3          )
4
5  # pull in common dependencies and additional i2c hardware support
6  target_link_libraries(mcp9808_i2c pico_stdlib hardware_i2c)
7
8  # create map/bin/hex file etc.
9  pico_add_extra_outputs(mcp9808_i2c)
10
11 # add url via pico_set_program_url
12 example_auto_set_url(mcp9808_i2c)
```

mcp9808_i2c.c

The example code.

Pico Examples: https://github.com/raspberrypi/pico-examples/blob/master/i2c/mcp9808_i2c/mcp9808_i2c.c

```c
/**
 * Copyright (c) 2020 Raspberry Pi (Trading) Ltd.
 *
 * SPDX-License-Identifier: BSD-3-Clause
 */

#include <stdio.h>
#include <string.h>
#include "pico/stdlib.h"
#include "pico/binary_info.h"
#include "hardware/i2c.h"

/* Example code to talk to a MCP9808 ±0.5°C Digital temperature Sensor

   This reads and writes to registers on the board.

   Connections on Raspberry Pi Pico board, other boards may vary.

   GPIO PICO_DEFAULT_I2C_SDA_PIN (On Pico this is GP4 (physical pin 6)) -> SDA on MCP9808
   board
   GPIO PICO_DEFAULT_I2C_SCK_PIN (On Pico this is GP5 (physcial pin 7)) -> SCL on MCP9808
   board
   Vsys (physical pin 39) -> VDD on MCP9808 board
   GND (physical pin 38)  -> GND on MCP9808 board

*/
//The bus address is determined by the state of pins A0, A1 and A2 on the MCP9808 board
static uint8_t ADDRESS = 0x18;

//hardware registers

const uint8_t REG_POINTER = 0x00;
const uint8_t REG_CONFIG = 0x01;
const uint8_t REG_TEMP_UPPER = 0x02;
const uint8_t REG_TEMP_LOWER = 0x03;
const uint8_t REG_TEMP_CRIT = 0x04;
const uint8_t REG_TEMP_AMB = 0x05;
const uint8_t REG_RESOLUTION = 0x08;


void mcp9808_check_limits(uint8_t upper_byte) {

    // Check flags and raise alerts accordingly
    if ((upper_byte & 0x40) == 0x40) { //TA > TUPPER
        printf("Temperature is above the upper temperature limit.\n");
```

```
 44      }
 45      if ((upper_byte & 0x20) == 0x20) { //TA < TLOWER
 46          printf("Temperature is below the lower temperature limit.\n");
 47      }
 48      if ((upper_byte & 0x80) == 0x80) { //TA > TCRIT
 49          printf("Temperature is above the critical temperature limit.\n");
 50      }
 51  }
 52
 53  float mcp9808_convert_temp(uint8_t upper_byte, uint8_t lower_byte) {
 54
 55      float temperature;
 56
 57
 58      //Check if TA <= 0°C and convert to denary accordingly
 59      if ((upper_byte & 0x10) == 0x10) {
 60          upper_byte = upper_byte & 0x0F;
 61          temperature = 256 - (((float) upper_byte * 16) + ((float) lower_byte / 16));
 62      } else {
 63          temperature = (((float) upper_byte * 16) + ((float) lower_byte / 16));
 64
 65      }
 66      return temperature;
 67  }
 68
 69  #ifdef i2c_default
 70  void mcp9808_set_limits() {
 71
 72      //Set an upper limit of 30°C for the temperature
 73      uint8_t upper_temp_msb = 0x01;
 74      uint8_t upper_temp_lsb = 0xE0;
 75
 76      //Set a lower limit of 20°C for the temperature
 77      uint8_t lower_temp_msb = 0x01;
 78      uint8_t lower_temp_lsb = 0x40;
 79
 80      //Set a critical limit of 40°C for the temperature
 81      uint8_t crit_temp_msb = 0x02;
 82      uint8_t crit_temp_lsb = 0x80;
 83
 84      uint8_t buf[3];
 85      buf[0] = REG_TEMP_UPPER;
 86      buf[1] = upper_temp_msb;
 87      buf[2] = upper_temp_lsb;
 88      i2c_write_blocking(i2c_default, ADDRESS, buf, 3, false);
 89
 90      buf[0] = REG_TEMP_LOWER;
 91      buf[1] = lower_temp_msb;
 92      buf[2] = lower_temp_lsb;
 93      i2c_write_blocking(i2c_default, ADDRESS, buf, 3, false);
 94
 95      buf[0] = REG_TEMP_CRIT;
 96      buf[1] = crit_temp_msb;
 97      buf[2] = crit_temp_lsb;;
 98      i2c_write_blocking(i2c_default, ADDRESS, buf, 3, false);
 99  }
100  #endif
101
102  int main() {
103
104      stdio_init_all();
105
106  #if !defined(i2c_default) || !defined(PICO_DEFAULT_I2C_SDA_PIN) ||
```

```
        !defined(PICO_DEFAULT_I2C_SCL_PIN)
107 #warning i2c/mcp9808_i2c example requires a board with I2C pins
108     puts("Default I2C pins were not defined");
109 #else
110     printf("Hello, MCP9808! Reading raw data from registers...\n");
111
112     // This example will use I2C0 on the default SDA and SCL pins (4, 5 on a Pico)
113     i2c_init(i2c_default, 400 * 1000);
114     gpio_set_function(PICO_DEFAULT_I2C_SDA_PIN, GPIO_FUNC_I2C);
115     gpio_set_function(PICO_DEFAULT_I2C_SCL_PIN, GPIO_FUNC_I2C);
116     gpio_pull_up(PICO_DEFAULT_I2C_SDA_PIN);
117     gpio_pull_up(PICO_DEFAULT_I2C_SCL_PIN);
118     // Make the I2C pins available to picotool
119     bi_decl(bi_2pins_with_func(PICO_DEFAULT_I2C_SDA_PIN, PICO_DEFAULT_I2C_SCL_PIN, GPIO_FUNC_I2C));
120
121     mcp9808_set_limits();
122
123     uint8_t buf[2];
124     uint16_t upper_byte;
125     uint16_t lower_byte;
126
127     float temperature;
128
129     while (1) {
130         // Start reading ambient temperature register for 2 bytes
131         i2c_write_blocking(i2c_default, ADDRESS, &REG_TEMP_AMB, 1, true);
132         i2c_read_blocking(i2c_default, ADDRESS, buf, 2, false);
133
134         upper_byte = buf[0];
135         lower_byte = buf[1];
136
137         //isolates limit flags in upper byte
138         mcp9808_check_limits(upper_byte & 0xE0);
139
140         //clears flag bits in upper byte
141         temperature = mcp9808_convert_temp(upper_byte & 0x1F, lower_byte);
142         printf("Ambient temperature: %.4f°C\n", temperature);
143
144         sleep_ms(1000);
145     }
146 #endif
147 }
```

Bill of Materials

Table 19. A list of materials required for the example

Item	Quantity	Details
Breadboard	1	generic part
Raspberry Pi Pico	1	https://www.raspberrypi.com/products/raspberry-pi-pico/
MCP9808 board	1	https://www.adafruit.com/product/1782
M/M Jumper wires	4	generic part

Attaching a MMA8451 3-axis digital accelerometer via I2C

This example code shows how to interface the Raspberry Pi Pico to the MMA8451 digital accelerometer sensor board.

This example reads and displays the acceleration values of the board in the 3 axis. It also allows the user to set the trade-off between the range and precision based on the values they require. Values often have an offset which can be accounted for by writing to the offset correction registers. The datasheet for the sensor can be found at https://cdn-shop.adafruit.com/datasheets/MMA8451Q-1.pdf for additional information.

Wiring information

Wiring up the device requires 4 jumpers, to connect VIN, GND, SDA and SCL. The example here uses I2C port 0, which is assigned to GPIO 4 (SDA) and 5 (SCL) in software. Power is supplied from the VSYS pin.

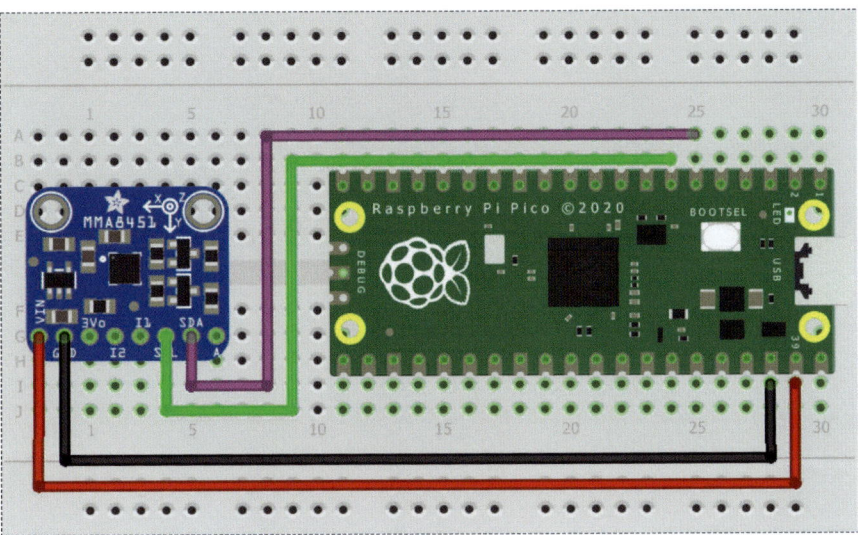

Figure 21. Wiring Diagram for MMA8451.

List of Files

CMakeLists.txt

CMake file to incorporate the example in to the examples build tree.

Pico Examples: https://github.com/raspberrypi/pico-examples/blob/master/i2c/mma8451_i2c/CMakeLists.txt

```
1  add_executable(mma8451_i2c
2          mma8451_i2c.c
3          )
4  # pull in common dependencies and additional i2c hardware support
5  target_link_libraries(mma8451_i2c pico_stdlib hardware_i2c)
6
7  # create map/bin/hex file etc.
8  pico_add_extra_outputs(mma8451_i2c)
9
10 # add url via pico_set_program_url
11 example_auto_set_url(mma8451_i2c)
```

mma8451_i2c.c

The example code.

Pico Examples: *https://github.com/raspberrypi/pico-examples/blob/master/i2c/mma8451_i2c/mma8451_i2c.c*

```c
/**
 * Copyright (c) 2020 Raspberry Pi (Trading) Ltd.
 *
 * SPDX-License-Identifier: BSD-3-Clause
 */

#include <stdio.h>
#include <string.h>
#include "pico/stdlib.h"
#include "pico/binary_info.h"
#include "hardware/i2c.h"

/* Example code to talk to a MMA8451 triple-axis accelerometer.

   This reads and writes to registers on the board.

   Connections on Raspberry Pi Pico board, other boards may vary.

   GPIO PICO_DEFAULT_I2C_SDA_PIN (On Pico this is GP4 (physical pin 6)) -> SDA on MMA8451 board
   GPIO PICO_DEFAULT_I2C_SCK_PIN (On Pico this is GP5 (physcial pin 7)) -> SCL on MMA8451 board
   VSYS (physical pin 39) -> VDD on MMA8451 board
   GND (physical pin 38)  -> GND on MMA8451 board

*/

const uint8_t ADDRESS = 0x1D;

//hardware registers

const uint8_t REG_X_MSB = 0x01;
const uint8_t REG_X_LSB = 0x02;
const uint8_t REG_Y_MSB = 0x03;
const uint8_t REG_Y_LSB = 0x04;
const uint8_t REG_Z_MSB = 0x05;
const uint8_t REG_Z_LSB = 0x06;
const uint8_t REG_DATA_CFG = 0x0E;
const uint8_t REG_CTRL_REG1 = 0x2A;

// Set the range and precision for the data
const uint8_t range_config = 0x01; // 0x00 for ±2g, 0x01 for ±4g, 0x02 for ±8g
const float count = 2048; // 4096 for ±2g, 2048 for ±4g, 1024 for ±8g

uint8_t buf[2];

float mma8451_convert_accel(uint16_t raw_accel) {
    float acceleration;
    // Acceleration is read as a multiple of g (gravitational acceleration on the Earth's surface)
    // Check if acceleration < 0 and convert to decimal accordingly
    if ((raw_accel & 0x2000) == 0x2000) {
        raw_accel &= 0x1FFF;
        acceleration = (-8192 + (float) raw_accel) / count;
    } else {
        acceleration = (float) raw_accel / count;
    }
    acceleration *= 9.81f;
    return acceleration;
}

```

```c
#ifdef i2c_default
void mma8451_set_state(uint8_t state) {
    buf[0] = REG_CTRL_REG1;
    buf[1] = state; // Set RST bit to 1
    i2c_write_blocking(i2c_default, ADDRESS, buf, 2, false);
}
#endif

int main() {
    stdio_init_all();

#if !defined(i2c_default) || !defined(PICO_DEFAULT_I2C_SDA_PIN) || !defined(PICO_DEFAULT_I2C_SCL_PIN)
#warning i2c/mma8451_i2c example requires a board with I2C pins
    puts("Default I2C pins were not defined");
#else
    printf("Hello, MMA8451! Reading raw data from registers...\n");

    // This example will use I2C0 on the default SDA and SCL pins (4, 5 on a Pico)
    i2c_init(i2c_default, 400 * 1000);
    gpio_set_function(PICO_DEFAULT_I2C_SDA_PIN, GPIO_FUNC_I2C);
    gpio_set_function(PICO_DEFAULT_I2C_SCL_PIN, GPIO_FUNC_I2C);
    gpio_pull_up(PICO_DEFAULT_I2C_SDA_PIN);
    gpio_pull_up(PICO_DEFAULT_I2C_SCL_PIN);
    // Make the I2C pins available to picotool
    bi_decl(bi_2pins_with_func(PICO_DEFAULT_I2C_SDA_PIN, PICO_DEFAULT_I2C_SCL_PIN, GPIO_FUNC_I2C));

    float x_acceleration;
    float y_acceleration;
    float z_acceleration;

    // Enable standby mode
    mma8451_set_state(0x00);

    // Edit configuration while in standby mode
    buf[0] = REG_DATA_CFG;
    buf[1] = range_config;
    i2c_write_blocking(i2c_default, ADDRESS, buf, 2, false);

    // Enable active mode
    mma8451_set_state(0x01);

    while (1) {

        // Start reading acceleration registers for 2 bytes
        i2c_write_blocking(i2c_default, ADDRESS, &REG_X_MSB, 1, true);
        i2c_read_blocking(i2c_default, ADDRESS, buf, 2, false);
        x_acceleration = mma8451_convert_accel(buf[0] << 6 | buf[1] >> 2);

        i2c_write_blocking(i2c_default, ADDRESS, &REG_Y_MSB, 1, true);
        i2c_read_blocking(i2c_default, ADDRESS, buf, 2, false);
        y_acceleration = mma8451_convert_accel(buf[0] << 6 | buf[1] >> 2);

        i2c_write_blocking(i2c_default, ADDRESS, &REG_Z_MSB, 1, true);
        i2c_read_blocking(i2c_default, ADDRESS, buf, 2, false);
        z_acceleration = mma8451_convert_accel(buf[0] << 6 | buf[1] >> 2);

        // Display acceleration values
        printf("ACCELERATION VALUES: \n");
        printf("X acceleration: %.6fms^-2\n", x_acceleration);
        printf("Y acceleration: %.6fms^-2\n", y_acceleration);
        printf("Z acceleration: %.6fms^-2\n", z_acceleration);
```

```
120
121        sleep_ms(500);
122
123        // Clear terminal
124        printf("\e[1;1H\e[2J");
125    }
126
127 #endif
128 }
```

Bill of Materials

Table 20. A list of materials required for the example

Item	Quantity	Details
Breadboard	1	generic part
Raspberry Pi Pico	1	https://www.raspberrypi.com/products/raspberry-pi-pico/
MMA8451 board	1	https://www.adafruit.com/product/2019
M/M Jumper wires	4	generic part

Attaching an MPL3115A2 altimeter via I2C

This example code shows how to interface the Raspberry Pi Pico to an MPL3115A2 altimeter via I2C. The MPL3115A2 has onboard pressure and temperature sensors which are used to estimate the altitude. In comparison to the BMP-family of pressure and temperature sensors, the MPL3115A2 has two interrupt pins for ultra low power operation and takes care of the sensor reading compensation on the board! It also has multiple modes of operation and impressive operating conditions.

The board used in this example comes from Adafruit, but any MPL3115A2 breakouts should work similarly.

The MPL3115A2 makes available two ways of reading its temperature and pressure data. The first is known as polling, where the Pico will continuously read data out of a set of auto-incrementing registers which are refreshed with new data every so often. The second, which this example will demonstrate, uses a 160-byte first-in-first-out (FIFO) queue and configurable interrupts to tell the Pico when to read data. More information regarding when the interrupts can be triggered available in the datasheet. This example waits for the 32 sample FIFO to overflow, detects this via an interrupt pin, and then averages the 32 samples taken. The sensor is configured to take a sample every second.

Bit math is used to convert the temperature and altitude data from the raw bits collected in the registers. Take the temperature calculation as an example: it is a 12-bit signed number with 8 integer bits and 4 fractional bits. First, we read the 2 8-bit registers and store them in a buffer. Then, we concatenate them into one unsigned 16-bit integer starting with the OUT_T_MSB register, thus making sure that the last bit of this register is aligned with the MSB in our 16 bit unsigned integer so it is correctly interpreted as the signed bit when we later cast this to a signed 16-bit integer. Finally, the entire number is converted to a float implicitly when we multiply it by 1/2^8 to shift it 8 bits to the right of the decimal point. Though only the last 4 bits of the OUT_T_LSB register hold data, this does not matter as the remaining 4 are held at zero and "disappear" when we shift the decimal point left by 8. Similar logic is applied to the altitude calculation.

> **TIP**
>
> Choosing the right sensor for your project among so many choices can be hard! There are multiple factors you may have to consider in addition to any constraints imposed on you. Cost, operating temperature, sensor resolution, power consumption, ease of use, communication protocols and supply voltage are all but a few factors that can play a role in sensor choice. For most hobbyist purposes though, the majority of sensors out there will do just fine!

Wiring information

Wiring up the device requires 5 jumpers, to connect VCC (3.3v), GND, INT1, SDA and SCL. The example here uses I2C port 0, which is assigned to GPIO 4 (SDA) and GPIO 5 (SCL) by default. Power is supplied from the 3.3V pin.

> **NOTE**
>
> The MPL3115A2 has a 1.6-3.6V voltage supply range. This means it can work with the Pico's 3.3v pins out of the box but our Adafruit breakout has an onboard voltage regulator for good measure. This may not always be true of other sensors, though.

Figure 22. Wiring Diagram for MPL3115A2 altimeter.

List of Files

CMakeLists.txt

CMake file to incorporate the example in to the examples build tree.

Pico Examples: https://github.com/raspberrypi/pico-examples/blob/master/i2c/mpl3115a2_i2c/CMakeLists.txt

```
1  add_executable(mpl3115a2_i2c
2          mpl3115a2_i2c.c
3          )
4
5  # pull in common dependencies and additional i2c hardware support
6  target_link_libraries(mpl3115a2_i2c pico_stdlib hardware_i2c)
7
8  # create map/bin/hex file etc.
9  pico_add_extra_outputs(mpl3115a2_i2c)
```

```
10
11 # add url via pico_set_program_url
12 example_auto_set_url(mpl3115a2_i2c)
```

mpl3115a2_i2c.c

The example code.

Pico Examples: https://github.com/raspberrypi/pico-examples/blob/master/i2c/mpl3115a2_i2c/mpl3115a2_i2c.c

```c
1  /**
2   * Copyright (c) 2021 Raspberry Pi (Trading) Ltd.
3   *
4   * SPDX-License-Identifier: BSD-3-Clause
5   */
6
7  #include <stdio.h>
8  #include "pico/stdlib.h"
9  #include "pico/binary_info.h"
10 #include "hardware/gpio.h"
11 #include "hardware/i2c.h"
12
13 /* Example code to talk to an MPL3115A2 altimeter sensor via I2C
14
15    See accompanying documentation in README.adoc or the C++ SDK booklet.
16
17    Connections on Raspberry Pi Pico board, other boards may vary.
18
19    GPIO PICO_DEFAULT_I2C_SDA_PIN (On Pico this is 4 (pin 6)) -> SDA on MPL3115A2 board
20    GPIO PICO_DEFAULT_I2C_SCK_PIN (On Pico this is 5 (pin 7)) -> SCL on MPL3115A2 board
21    GPIO 16 -> INT1 on MPL3115A2 board
22    3.3v (pin 36) -> VCC on MPL3115A2 board
23    GND (pin 38)  -> GND on MPL3115A2 board
24 */
25
26 // 7-bit address
27 #define ADDR 0x60
28 #define INT1_PIN _u(16)
29
30 // following definitions only valid for F_MODE > 0 (ie. if FIFO enabled)
31 #define MPL3115A2_F_DATA _u(0x01)
32 #define MPL3115A2_F_STATUS _u(0x00)
33 #define MPL3115A2_F_SETUP _u(0x0F)
34 #define MPL3115A2_INT_SOURCE _u(0x12)
35 #define MPL3115A2_CTRLREG1 _u(0x26)
36 #define MPL3115A2_CTRLREG2 _u(0x27)
37 #define MPL3115A2_CTRLREG3 _u(0x28)
38 #define MPL3115A2_CTRLREG4 _u(0x29)
39 #define MPL3115A2_CTRLREG5 _u(0x2A)
40 #define MPL3115A2_PT_DATA_CFG _u(0x13)
41 #define MPL3115A2_OFF_P _u(0x2B)
42 #define MPL3115A2_OFF_T _u(0x2C)
43 #define MPL3115A2_OFF_H _u(0x2D)
44
45 #define MPL3115A2_FIFO_DISABLED _u(0x00)
46 #define MPL3115A2_FIFO_STOP_ON_OVERFLOW _u(0x80)
47 #define MPL3115A2_FIFO_SIZE 32
48 #define MPL3115A2_DATA_BATCH_SIZE 5
49 #define MPL3115A2_ALTITUDE_NUM_REGS 3
50 #define MPL3115A2_ALTITUDE_INT_SIZE 20
51 #define MPL3115A2_TEMPERATURE_INT_SIZE 12
52 #define MPL3115A2_NUM_FRAC_BITS 4
```

```c
53
54  #define PARAM_ASSERTIONS_ENABLE_I2C 1
55
56  volatile uint8_t fifo_data[MPL3115A2_FIFO_SIZE * MPL3115A2_DATA_BATCH_SIZE];
57  volatile bool has_new_data = false;
58
59  struct mpl3115a2_data_t {
60      // Q8.4 fixed point
61      float temperature;
62      // Q16.4 fixed-point
63      float altitude;
64  };
65
66  void copy_to_vbuf(uint8_t buf1[], volatile uint8_t buf2[], int buflen) {
67      for (size_t i = 0; i < buflen; i++) {
68          buf2[i] = buf1[i];
69      }
70  }
71
72  #ifdef i2c_default
73
74  void mpl3115a2_read_fifo(volatile uint8_t fifo_buf[]) {
75      // drains the 160 byte FIFO
76      uint8_t reg = MPL3115A2_F_DATA;
77      uint8_t buf[MPL3115A2_FIFO_SIZE * MPL3115A2_DATA_BATCH_SIZE];
78      i2c_write_blocking(i2c_default, ADDR, &reg, 1, true);
79      // burst read 160 bytes from fifo
80      i2c_read_blocking(i2c_default, ADDR, buf, MPL3115A2_FIFO_SIZE * MPL3115A2_DATA_BATCH_SIZE, false);
81      copy_to_vbuf(buf, fifo_buf, MPL3115A2_FIFO_SIZE * MPL3115A2_DATA_BATCH_SIZE);
82  }
83
84  uint8_t mpl3115a2_read_reg(uint8_t reg) {
85      uint8_t read;
86      i2c_write_blocking(i2c_default, ADDR, &reg, 1, true); // keep control of bus
87      i2c_read_blocking(i2c_default, ADDR, &read, 1, false);
88      return read;
89  }
90
91  void mpl3115a2_init() {
92      // set as altimeter with oversampling ratio of 128
93      uint8_t buf[] = {MPL3115A2_CTRLREG1, 0xB8};
94      i2c_write_blocking(i2c_default, ADDR, buf, 2, false);
95
96      // set data refresh every 2 seconds, 0 next bits as we're not using those interrupts
97      buf[0] = MPL3115A2_CTRLREG2, buf[1] = 0x00;
98      i2c_write_blocking(i2c_default, ADDR, buf, 2, false);
99
100     // set both interrupts pins to active low and enable internal pullups
101     buf[0] = MPL3115A2_CTRLREG3, buf[1] = 0x01;
102     i2c_write_blocking(i2c_default, ADDR, buf, 2, false);
103
104     // enable FIFO interrupt
105     buf[0] = MPL3115A2_CTRLREG4, buf[1] = 0x40;
106     i2c_write_blocking(i2c_default, ADDR, buf, 2, false);
107
108     // tie FIFO interrupt to pin INT1
109     buf[0] = MPL3115A2_CTRLREG5, buf[1] = 0x40;
110     i2c_write_blocking(i2c_default, ADDR, buf, 2, false);
111
112     // set p, t and h offsets here if needed
113     // eg. 2's complement number: 0xFF subtracts 1 meter
114     //buf[0] = MPL3115A2_OFF_H, buf[1] = 0xFF;
```

```c
115        //i2c_write_blocking(i2c_default, ADDR, buf, 2, false);
116
117        // do not accept more data on FIFO overflow
118        buf[0] = MPL3115A2_F_SETUP, buf[1] = MPL3115A2_FIFO_STOP_ON_OVERFLOW;
119        i2c_write_blocking(i2c_default, ADDR, buf, 2, false);
120
121        // set device active
122        buf[0] = MPL3115A2_CTRLREG1, buf[1] = 0xB9;
123        i2c_write_blocking(i2c_default, ADDR, buf, 2, false);
124 }
125
126 void gpio_callback(uint gpio, uint32_t events) {
127        // if we had enabled more than 2 interrupts on same pin, then we should read
128        // INT_SOURCE reg to find out which interrupt triggered
129
130        // we can filter by which GPIO was triggered
131        if (gpio == INT1_PIN) {
132            // FIFO overflow interrupt
133            // watermark bits set to 0 in F_SETUP reg, so only possible event is an overflow
134            // otherwise, we would read F_STATUS to confirm it was an overflow
135            printf("FIFO overflow!\n");
136            // drain the fifo
137            mpl3115a2_read_fifo(fifo_data);
138            // read status register to clear interrupt bit
139            mpl3115a2_read_reg(MPL3115A2_F_STATUS);
140            has_new_data = true;
141        }
142 }
143
144 #endif
145
146 void mpl3115a2_convert_fifo_batch(uint8_t start, volatile uint8_t buf[], struct mpl3115a2_data_t *data) {
147        // convert a batch of fifo data into temperature and altitude data
148
149        // 3 altitude registers: MSB (8 bits), CSB (8 bits) and LSB (4 bits, starting from MSB)
150        // first two are integer bits (2's complement) and LSB is fractional bits -> makes 20 bit signed integer
151        int32_t h = (int32_t) ((uint32_t) buf[start] << 24 | buf[start + 1] << 16 | buf[start + 2] << 8);
152        data->altitude = ((float)h) / 65536.f;
153
154        // 2 temperature registers: MSB (8 bits) and LSB (4 bits, starting from MSB)
155        // first 8 are integer bits with sign and LSB is fractional bits -> 12 bit signed integer
156        int16_t t = (int16_t) (((uint16_t) buf[start + 3]) << 8 | buf[start + 4]);
157        data->temperature = ((float)t) / 256.f;
158 }
159
160 int main() {
161        stdio_init_all();
162 #if !defined(i2c_default) || !defined(PICO_DEFAULT_I2C_SDA_PIN) || !defined(PICO_DEFAULT_I2C_SCL_PIN)
163 #warning i2c / mpl3115a2_i2c example requires a board with I2C pins
164        puts("Default I2C pins were not defined");
165 #else
166        printf("Hello, MPL3115A2. Waiting for something to interrupt me!...\n");
167
168        // use default I2C0 at 400kHz, I2C is active low
169        i2c_init(i2c_default, 400 * 1000);
170        gpio_set_function(PICO_DEFAULT_I2C_SDA_PIN, GPIO_FUNC_I2C);
171        gpio_set_function(PICO_DEFAULT_I2C_SCL_PIN, GPIO_FUNC_I2C);
172        gpio_pull_up(PICO_DEFAULT_I2C_SDA_PIN);
173        gpio_pull_up(PICO_DEFAULT_I2C_SCL_PIN);
```

```
174
175        gpio_init(INT1_PIN);
176        gpio_pull_up(INT1_PIN); // pull it up even more!
177
178        // add program information for picotool
179        bi_decl(bi_program_name("Example in the pico-examples library for the MPL3115A2
    altimeter"));
180        bi_decl(bi_1pin_with_name(16, "Interrupt pin 1"));
181        bi_decl(bi_2pins_with_func(PICO_DEFAULT_I2C_SDA_PIN, PICO_DEFAULT_I2C_SCL_PIN,
    GPIO_FUNC_I2C));
182
183        mpl3115a2_init();
184
185        gpio_set_irq_enabled_with_callback(INT1_PIN, GPIO_IRQ_LEVEL_LOW, true, &gpio_callback);
186
187        while (1) {
188            // as interrupt data comes in, let's print the 32 sample average
189            if (has_new_data) {
190                float tsum = 0, hsum = 0;
191                struct mpl3115a2_data_t data;
192                for (int i = 0; i < MPL3115A2_FIFO_SIZE; i++) {
193                    mpl3115a2_convert_fifo_batch(i * MPL3115A2_DATA_BATCH_SIZE, fifo_data,
    &data);
194                    tsum += data.temperature;
195                    hsum += data.altitude;
196                }
197                printf("%d sample average -> t: %.4f C, h: %.4f m\n", MPL3115A2_FIFO_SIZE, tsum
    / MPL3115A2_FIFO_SIZE,
198                       hsum / MPL3115A2_FIFO_SIZE);
199                has_new_data = false;
200            }
201            sleep_ms(10);
202        };
203
204 #endif
205        return 0;
206 }
```

Bill of Materials

Table 21. A list of materials required for the example

Item	Quantity	Details
Breadboard	1	generic part
Raspberry Pi Pico	1	https://www.raspberrypi.com/products/raspberry-pi-pico/
MPL3115A2 altimeter	1	Adafruit
M/M Jumper wires	5	generic part

Attaching an OLED display via I2C

This example code shows how to interface the Raspberry Pi Pico with an 128x32 OLED display board based on the SSD1306 display driver, datasheet here.

The code displays a series of tiny raspberries that scroll horizontally, in the process showing you how to initialize the display, write to the entire display, write to only a portion of the display, and configure scrolling.

The SSD1306 is operated via a list of versatile commands (see datasheet) that allows the user to access all the capabilities of the driver. After sending a slave address, the data that follows can be either a command, flags to follow up a command or data to be written directly into the display's RAM. A control byte is required for each write after the slave address so that the driver knows what type of data is being sent.

This display is 32 pixels high by 128 pixels wide. These 32 vertical pixels are partitioned into 4 pages, each 8 pixels in height. In RAM, this looks roughly like:

> **NOTE**
>
> The SSD1306 can drive displays that are up to 64 pixels high and 128 pixels wide.

```
         | COL0 | COL1 | COL2 | COL3 |  ...  | COL126 | COL127 |
  PAGE 0 |      |      |      |      |       |        |        |
  PAGE 1 |      |      |      |      |       |        |        |
  PAGE 2 |      |      |      |      |       |        |        |
  PAGE 3 |      |      |      |      |       |        |        |
  ----------------------------------------------------------------
```

Within each page, we have:

```
         | COL0 | COL1 | COL2 | COL3 |  ...  | COL126 | COL127 |
  COM 0  |      |      |      |      |       |        |        |
  COM 1  |      |      |      |      |       |        |        |
    :    |      |      |      |      |       |        |        |
  COM 7  |      |      |      |      |       |        |        |
  ----------------------------------------------------------------
```

> **NOTE**
>
> There is a difference between columns in RAM and the actual segment pads that connect the driver to the display. The RAM addresses COL0 - COL127 are mapped to these segment pins SEG0 - SEG127 by default. The distinction between these two is important as we can for example, easily mirror contents of RAM without rewriting a buffer.

The driver has 3 modes of transferring the pixels in RAM to the display (provided that the driver is set to use its RAM content to drive the display, ie. command 0xA4 is sent). We choose horizontal addressing mode which, after setting the column address and page address registers to our desired start positions, will increment the column address register until the OLED display width is reached (127 in our case) after which the column address register will reset to its starting value and the page address is incremented. Once the page register reaches the end, it will wrap around as well. Effectively, this scans across the display from top to bottom, left to right in blocks that are 8 pixels high. When a byte is sent to be written into RAM, it sets all the rows for the current position of the column address register. So, if we send 10101010, and we are on PAGE 0 and COL1, COM0 is set to 1, COM1 is set to 0, COM2 is set to 1, and so on. Effectively, the byte is "transposed" to fill a single page's column. The datasheet has further information on this and the two other modes.

Horizontal addressing mode has the key advantage that we can keep one single 512 byte buffer (128 columns x 4 pages and each byte fills a page's rows) and write this in one go to the RAM (column address auto increments on writes as well as reads) instead of working with 2D matrices of pixels and adding more overhead.

> **NOTE**
>
> - The SSD1306 is able to drive 128x64 displays but as our display is 128x32, only half of the COM (common) pins are connected to the display.
>
> - The specific display model being used is UG-2832HSWEG02

Wiring information

Wiring up the device requires 4 jumpers, to connect VCC (3.3v), GND, SDA and SCL and optionally a 5th jumper for the driver RESET pin. The example here uses the default I2C port 0, which is assigned to GPIO 4 (SDA) and 5 (SCL) in software. Power is supplied from the 3.3V pin from the Pico.

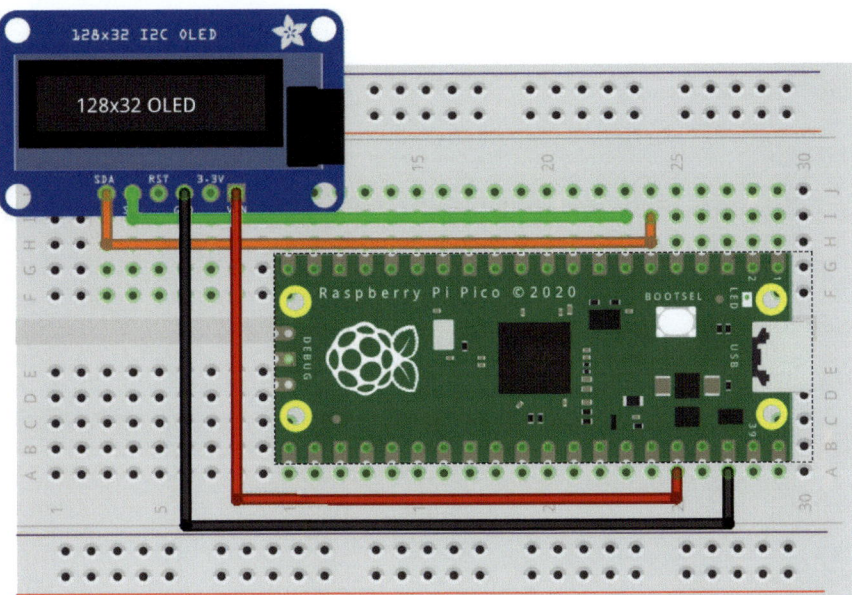

Figure 23. Wiring Diagram for oled display via I2C.

List of Files

CMakeLists.txt

CMake file to incorporate the example into the examples build tree.

Pico Examples: https://github.com/raspberrypi/pico-examples/blob/master/i2c/oled_i2c/CMakeLists.txt

```
1  add_executable(oled_i2c
2          oled_i2c.c
3          )
4
5  # pull in common dependencies and additional i2c hardware support
6  target_link_libraries(oled_i2c pico_stdlib hardware_i2c)
7
8  # create map/bin/hex file etc.
9  pico_add_extra_outputs(oled_i2c)
10
11 # add url via pico_set_program_url
12 example_auto_set_url(oled_i2c)
```

oled_i2c.c

The example code.

Pico Examples: *https://github.com/raspberrypi/pico-examples/blob/master/i2c/oled_i2c/oled_i2c.c*

```c
/**
 * Copyright (c) 2021 Raspberry Pi (Trading) Ltd.
 *
 * SPDX-License-Identifier: BSD-3-Clause
 */

#include <stdio.h>
#include <string.h>
#include <stdlib.h>
#include "pico/stdlib.h"
#include "pico/binary_info.h"
#include "hardware/i2c.h"
#include "raspberry26x32.h"

/* Example code to talk to an SSD1306-based OLED display

   NOTE: Ensure the device is capable of being driven at 3.3v NOT 5v. The Pico
   GPIO (and therefore I2C) cannot be used at 5v.

   You will need to use a level shifter on the I2C lines if you want to run the
   board at 5v.

   Connections on Raspberry Pi Pico board, other boards may vary.

   GPIO PICO_DEFAULT_I2C_SDA_PIN (on Pico this is GP4 (pin 6)) -> SDA on display
   board
   GPIO PICO_DEFAULT_I2C_SCK_PIN (on Pico this is GP5 (pin 7)) -> SCL on
   display board
   3.3v (pin 36) -> VCC on display board
   GND (pin 38)  -> GND on display board
*/

// commands (see datasheet)
#define OLED_SET_CONTRAST _u(0x81)
#define OLED_SET_ENTIRE_ON _u(0xA4)
#define OLED_SET_NORM_INV _u(0xA6)
#define OLED_SET_DISP _u(0xAE)
#define OLED_SET_MEM_ADDR _u(0x20)
#define OLED_SET_COL_ADDR _u(0x21)
#define OLED_SET_PAGE_ADDR _u(0x22)
#define OLED_SET_DISP_START_LINE _u(0x40)
#define OLED_SET_SEG_REMAP _u(0xA0)
#define OLED_SET_MUX_RATIO _u(0xA8)
#define OLED_SET_COM_OUT_DIR _u(0xC0)
#define OLED_SET_DISP_OFFSET _u(0xD3)
#define OLED_SET_COM_PIN_CFG _u(0xDA)
#define OLED_SET_DISP_CLK_DIV _u(0xD5)
#define OLED_SET_PRECHARGE _u(0xD9)
#define OLED_SET_VCOM_DESEL _u(0xDB)
#define OLED_SET_CHARGE_PUMP _u(0x8D)
#define OLED_SET_HORIZ_SCROLL _u(0x26)
#define OLED_SET_SCROLL _u(0x2E)

#define OLED_ADDR _u(0x3C)
#define OLED_HEIGHT _u(32)
#define OLED_WIDTH _u(128)
#define OLED_PAGE_HEIGHT _u(8)
```

```c
58 #define OLED_NUM_PAGES OLED_HEIGHT / OLED_PAGE_HEIGHT
59 #define OLED_BUF_LEN (OLED_NUM_PAGES * OLED_WIDTH)
60
61 #define OLED_WRITE_MODE _u(0xFE)
62 #define OLED_READ_MODE _u(0xFF)
63
64 struct render_area {
65     uint8_t start_col;
66     uint8_t end_col;
67     uint8_t start_page;
68     uint8_t end_page;
69
70     int buflen;
71 };
72
73 void fill(uint8_t buf[], uint8_t fill) {
74     // fill entire buffer with the same byte
75     for (int i = 0; i < OLED_BUF_LEN; i++) {
76         buf[i] = fill;
77     }
78 };
79
80 void fill_page(uint8_t *buf, uint8_t fill, uint8_t page) {
81     // fill entire page with the same byte
82     memset(buf + (page * OLED_WIDTH), fill, OLED_WIDTH);
83 };
84
85 // convenience methods for printing out a buffer to be rendered
86 // mostly useful for debugging images, patterns, etc
87
88 void print_buf_page(uint8_t buf[], uint8_t page) {
89     // prints one page of a full length (128x4) buffer
90     for (int j = 0; j < OLED_PAGE_HEIGHT; j++) {
91         for (int k = 0; k < OLED_WIDTH; k++) {
92             printf("%u", (buf[page * OLED_WIDTH + k] >> j) & 0x01);
93         }
94         printf("\n");
95     }
96 }
97
98 void print_buf_pages(uint8_t buf[]) {
99     // prints all pages of a full length buffer
100     for (int i = 0; i < OLED_NUM_PAGES; i++) {
101         printf("--page %d--\n", i);
102         print_buf_page(buf, i);
103     }
104 }
105
106 void print_buf_area(uint8_t *buf, struct render_area *area) {
107     // print a render area of generic size
108     int area_width = area->end_col - area->start_col + 1;
109     int area_height = area->end_page - area->start_page + 1; // in pages, not pixels
110     for (int i = 0; i < area_height; i++) {
111         for (int j = 0; j < OLED_PAGE_HEIGHT; j++) {
112             for (int k = 0; k < area_width; k++) {
113                 printf("%u", (buf[i * area_width + k] >> j) & 0x01);
114             }
115             printf("\n");
116         }
117     }
118 }
119
120 void calc_render_area_buflen(struct render_area *area) {
```

```c
121      // calculate how long the flattened buffer will be for a render area
122      area->buflen = (area->end_col - area->start_col + 1) * (area->end_page - area->start_page + 1);
123 }
124
125 #ifdef i2c_default
126
127 void oled_send_cmd(uint8_t cmd) {
128      // I2C write process expects a control byte followed by data
129      // this "data" can be a command or data to follow up a command
130
131      // Co = 1, D/C = 0 => the driver expects a command
132      uint8_t buf[2] = {0x80, cmd};
133      i2c_write_blocking(i2c_default, (OLED_ADDR & OLED_WRITE_MODE), buf, 2, false);
134 }
135
136 void oled_send_buf(uint8_t buf[], int buflen) {
137      // in horizontal addressing mode, the column address pointer auto-increments
138      // and then wraps around to the next page, so we can send the entire frame
139      // buffer in one gooooooo!
140
141      // copy our frame buffer into a new buffer because we need to add the control byte
142      // to the beginning
143
144      // TODO find a more memory-efficient way to do this..
145      // maybe break the data transfer into pages?
146      uint8_t *temp_buf = malloc(buflen + 1);
147
148      for (int i = 1; i < buflen + 1; i++) {
149          temp_buf[i] = buf[i - 1];
150      }
151      // Co = 0, D/C = 1 => the driver expects data to be written to RAM
152      temp_buf[0] = 0x40;
153      i2c_write_blocking(i2c_default, (OLED_ADDR & OLED_WRITE_MODE), temp_buf, buflen + 1, false);
154
155      free(temp_buf);
156 }
157
158 void oled_init() {
159      // some of these commands are not strictly necessary as the reset
160      // process defaults to some of these but they are shown here
161      // to demonstrate what the initialization sequence looks like
162
163      // some configuration values are recommended by the board manufacturer
164
165      oled_send_cmd(OLED_SET_DISP | 0x00); // set display off
166
167      /* memory mapping */
168      oled_send_cmd(OLED_SET_MEM_ADDR); // set memory address mode
169      oled_send_cmd(0x00); // horizontal addressing mode
170
171      /* resolution and layout */
172      oled_send_cmd(OLED_SET_DISP_START_LINE); // set display start line to 0
173
174      oled_send_cmd(OLED_SET_SEG_REMAP | 0x01); // set segment re-map
175      // column address 127 is mapped to SEG0
176
177      oled_send_cmd(OLED_SET_MUX_RATIO); // set multiplex ratio
178      oled_send_cmd(OLED_HEIGHT - 1); // our display is only 32 pixels high
179
180      oled_send_cmd(OLED_SET_COM_OUT_DIR | 0x08); // set COM (common) output scan direction
181      // scan from bottom up, COM[N-1] to COM0
```

```c
182
183        oled_send_cmd(OLED_SET_DISP_OFFSET); // set display offset
184        oled_send_cmd(0x00); // no offset
185
186        oled_send_cmd(OLED_SET_COM_PIN_CFG); // set COM (common) pins hardware configuration
187        oled_send_cmd(0x02); // manufacturer magic number
188
189        /* timing and driving scheme */
190        oled_send_cmd(OLED_SET_DISP_CLK_DIV); // set display clock divide ratio
191        oled_send_cmd(0x80); // div ratio of 1, standard freq
192
193        oled_send_cmd(OLED_SET_PRECHARGE); // set pre-charge period
194        oled_send_cmd(0xF1); // Vcc internally generated on our board
195
196        oled_send_cmd(OLED_SET_VCOM_DESEL); // set VCOMH deselect level
197        oled_send_cmd(0x30); // 0.83xVcc
198
199        /* display */
200        oled_send_cmd(OLED_SET_CONTRAST); // set contrast control
201        oled_send_cmd(0xFF);
202
203        oled_send_cmd(OLED_SET_ENTIRE_ON); // set entire display on to follow RAM content
204
205        oled_send_cmd(OLED_SET_NORM_INV); // set normal (not inverted) display
206
207        oled_send_cmd(OLED_SET_CHARGE_PUMP); // set charge pump
208        oled_send_cmd(0x14); // Vcc internally generated on our board
209
210        oled_send_cmd(OLED_SET_SCROLL | 0x00); // deactivate horizontal scrolling if set
211        // this is necessary as memory writes will corrupt if scrolling was enabled
212
213        oled_send_cmd(OLED_SET_DISP | 0x01); // turn display on
214    }
215
216    void render(uint8_t *buf, struct render_area *area) {
217        // update a portion of the display with a render area
218        oled_send_cmd(OLED_SET_COL_ADDR);
219        oled_send_cmd(area->start_col);
220        oled_send_cmd(area->end_col);
221
222        oled_send_cmd(OLED_SET_PAGE_ADDR);
223        oled_send_cmd(area->start_page);
224        oled_send_cmd(area->end_page);
225
226        oled_send_buf(buf, area->buflen);
227    }
228
229    #endif
230
231    int main() {
232        stdio_init_all();
233
234    #if !defined(i2c_default) || !defined(PICO_DEFAULT_I2C_SDA_PIN) || !defined(PICO_DEFAULT_I2C_SCL_PIN)
235    #warning i2c / oled_i2d example requires a board with I2C pins
236        puts("Default I2C pins were not defined");
237    #else
238        // useful information for picotool
239        bi_decl(bi_2pins_with_func(PICO_DEFAULT_I2C_SDA_PIN, PICO_DEFAULT_I2C_SCL_PIN, GPIO_FUNC_I2C));
240        bi_decl(bi_program_description("OLED I2C example for the Raspberry Pi Pico"));
241
242        printf("Hello, OLED display! Look at my raspberries..\n");
```

```c
    // I2C is "open drain", pull ups to keep signal high when no data is being
    // sent
    i2c_init(i2c_default, 400 * 1000);
    gpio_set_function(PICO_DEFAULT_I2C_SDA_PIN, GPIO_FUNC_I2C);
    gpio_set_function(PICO_DEFAULT_I2C_SCL_PIN, GPIO_FUNC_I2C);
    gpio_pull_up(PICO_DEFAULT_I2C_SDA_PIN);
    gpio_pull_up(PICO_DEFAULT_I2C_SCL_PIN);

    // run through the complete initialization process
    oled_init();

    // initialize render area for entire frame (128 pixels by 4 pages)
    struct render_area frame_area = {start_col: 0, end_col : OLED_WIDTH - 1, start_page : 0, end_page : OLED_NUM_PAGES - 1};
    calc_render_area_buflen(&frame_area);

    // zero the entire display
    uint8_t buf[OLED_BUF_LEN];
    fill(buf, 0x00);
    render(buf, &frame_area);

    // intro sequence: flash the screen 3 times
    for (int i = 0; i < 3; i++) {
        oled_send_cmd(0xA5); // ignore RAM, all pixels on
        sleep_ms(500);
        oled_send_cmd(0xA4); // go back to following RAM
        sleep_ms(500);
    }

    // render 3 cute little raspberries
    struct render_area area = {start_col: 0, end_col : IMG_WIDTH - 1, start_page : 0, end_page : OLED_NUM_PAGES - 1};
    calc_render_area_buflen(&area);
    render(raspberry26x32, &area);
    for (int i = 1; i < 3; i++) {
        uint8_t offset = 5 + IMG_WIDTH; // 5px padding
        area.start_col += offset;
        area.end_col += offset;
        render(raspberry26x32, &area);
    }

    // configure horizontal scrolling
    oled_send_cmd(OLED_SET_HORIZ_SCROLL | 0x00);
    oled_send_cmd(0x00); // dummy byte
    oled_send_cmd(0x00); // start page 0
    oled_send_cmd(0x00); // time interval
    oled_send_cmd(0x03); // end page 3
    oled_send_cmd(0x00); // dummy byte
    oled_send_cmd(0xFF); // dummy byte

    // let's goooo!
    oled_send_cmd(OLED_SET_SCROLL | 0x01);

#endif
    return 0;
}
```

Bill of Materials

Table 22. A list of materials required for the example

Item	Quantity	Details
Breadboard	1	generic part
Raspberry Pi Pico	1	https://www.raspberrypi.com/products/raspberry-pi-pico/
SSD1306-based OLED display	1	Adafruit part
M/M Jumper wires	4	generic part

Attaching a PA1010D Mini GPS module via I2C

This example code shows how to interface the Raspberry Pi Pico to the PA1010D Mini GPS module

This allows you to read basic location and time data from the Recommended Minimum Specific GNSS Sentence (GNRMC protocol) and displays it in a user-friendly format. The datasheet for the module can be found on https://cdn-learn.adafruit.com/assets/assets/000/084/295/original/CD_PA1010D_Datasheet_v.03.pdf?1573833002. The output sentence is read and parsed to split the data fields into a 2D character array, which are then individually printed out. The commands to use different protocols and change settings are found on https://www.sparkfun.com/datasheets/GPS/Modules/PMTK_Protocol.pdf. Additional protocols can be used by editing the `init_command` array.

> **ℹ NOTE**
>
> Each command requires a checksum after the asterisk. The checksum can be calculated for your command using the following website: https://nmeachecksum.eqth.net/.
>
> The GPS needs to be used outdoors in open skies and requires about 15 seconds to acquire a satellite signal in order to display valid data. When the signal is detected, the device will blink a green LED at 1 Hz.

Wiring information

Wiring up the device requires 4 jumpers, to connect VDD, GND, SDA and SCL. The example here uses I2C port 0, which is assigned to GPIO 4 (SDA) and 5 (SCL) in software. Power is supplied from the 3V pin.

Figure 24. Wiring Diagram for PA1010D.

List of Files

CMakeLists.txt

CMake file to incorporate the example in to the examples build tree.

Pico Examples: https://github.com/raspberrypi/pico-examples/blob/master/i2c/pa1010d_i2c/CMakeLists.txt

```
1  add_executable(pa1010d_i2c
2          pa1010d_i2c.c
3          )
4
5  # pull in common dependencies and additional i2c hardware support
6  target_link_libraries(pa1010d_i2c pico_stdlib hardware_i2c)
7
8  # create map/bin/hex file etc.
9  pico_add_extra_outputs(pa1010d_i2c)
10
11 # add url via pico_set_program_url
12 example_auto_set_url(pa1010d_i2c)
```

pa1010d_i2c.c

The example code.

Pico Examples: https://github.com/raspberrypi/pico-examples/blob/master/i2c/pa1010d_i2c/pa1010d_i2c.c

```
1  /**
2   * Copyright (c) 2020 Raspberry Pi (Trading) Ltd.
3   *
4   * SPDX-License-Identifier: BSD-3-Clause
5   */
6
7  #include <stdio.h>
8  #include <string.h>
9  #include "pico/stdlib.h"
10 #include "pico/binary_info.h"
11 #include "hardware/i2c.h"
12 #include "string.h"
13
14 /* Example code to talk to a PA1010D Mini GPS module.
15
16    This example reads the Recommended Minimum Specific GNSS Sentence, which includes basic
   location and time data, each second, formats and displays it.
17
18    Connections on Raspberry Pi Pico board, other boards may vary.
19
20    GPIO PICO_DEFAULT_I2C_SDA_PIN (On Pico this is 4 (physical pin 6)) -> SDA on PA1010D board
21    GPIO PICO_DEFAULT_I2C_SCK_PIN (On Pico this is 5 (physical pin 7)) -> SCL on PA1010D board
22    3.3v (physical pin 36) -> VCC on PA1010D board
23    GND (physical pin 38)  -> GND on PA1010D board
24 */
25
26 const int addr = 0x10;
27 const int max_read = 250;
28
29 #ifdef i2c_default
30
31 void pa1010d_write_command(const char command[], int com_length) {
32     // Convert character array to bytes for writing
33     uint8_t int_command[com_length];
34
```

```c
        for (int i = 0; i < com_length; ++i) {
            int_command[i] = command[i];
            i2c_write_blocking(i2c_default, addr, &int_command[i], 1, true);
        }
}

void pa1010d_parse_string(char output[], char protocol[]) {
    // Finds location of protocol message in output
    char *com_index = strstr(output, protocol);
    int p = com_index - output;

    // Splits components of output sentence into array
    int no_of_fields = 14;
    int max_len = 15;

    int n = 0;
    int m = 0;

    char gps_data[no_of_fields][max_len];
    memset(gps_data, 0, sizeof(gps_data));

    bool complete = false;
    while (output[p] != '$' && n < max_len && complete == false) {
        if (output[p] == ',' || output[p] == '*') {
            n += 1;
            m = 0;
        } else {
            gps_data[n][m] = output[p];
            // Checks if sentence is complete
            if (m < no_of_fields) {
                m++;
            } else {
                complete = true;
            }
        }
        p++;
    }

    // Displays GNRMC data
    // Similarly, additional if statements can be used to add more protocols
    if (strcmp(protocol, "GNRMC") == 0) {
        printf("Protcol:%s\n", gps_data[0]);
        printf("UTC Time: %s\n", gps_data[1]);
        printf("Status: %s\n", gps_data[2][0] == 'V' ? "Data invalid. GPS fix not found." : "Data Valid");
        printf("Latitude: %s\n", gps_data[3]);
        printf("N/S indicator: %s\n", gps_data[4]);
        printf("Longitude: %s\n", gps_data[5]);
        printf("E/W indicator: %s\n", gps_data[6]);
        printf("Speed over ground: %s\n", gps_data[7]);
        printf("Course over ground: %s\n", gps_data[8]);
        printf("Date: %c%c/%c%c/%c%c\n", gps_data[9][0], gps_data[9][1], gps_data[9][2],
               gps_data[9][3], gps_data[9][4],
               gps_data[9][5]);
        printf("Magnetic Variation: %s\n", gps_data[10]);
        printf("E/W degree indicator: %s\n", gps_data[11]);
        printf("Mode: %s\n", gps_data[12]);
        printf("Checksum: %c%c\n", gps_data[13][0], gps_data[13][1]);
    }
}

void pa1010d_read_raw(char numcommand[]) {
    uint8_t buffer[max_read];
```

```c
    int i = 0;
    bool complete = false;

    i2c_read_blocking(i2c_default, addr, buffer, max_read, false);

    // Convert bytes to characters
    while (i < max_read && complete == false) {
        numcommand[i] = buffer[i];
        // Stop converting at end of message
        if (buffer[i] == 10 && buffer[i + 1] == 10) {
            complete = true;
        }
        i++;
    }
}
#endif

int main() {
    stdio_init_all();
#if !defined(i2c_default) || !defined(PICO_DEFAULT_I2C_SDA_PIN) || !defined(PICO_DEFAULT_I2C_SCL_PIN)
#warning i2c/mpu6050_i2c example requires a board with I2C pins
    puts("Default I2C pins were not defined");
#else

    char numcommand[max_read];

    // Decide which protocols you would like to retrieve data from
    char init_command[] = "$PMTK314,0,1,0,0,0,0,0,0,0,0,0,0,0,0,0,0,0,0,0*29\r\n";

    // This example will use I2C0 on the default SDA and SCL pins (4, 5 on a Pico)
    i2c_init(i2c_default, 400 * 1000);
    gpio_set_function(PICO_DEFAULT_I2C_SDA_PIN, GPIO_FUNC_I2C);
    gpio_set_function(PICO_DEFAULT_I2C_SCL_PIN, GPIO_FUNC_I2C);
    gpio_pull_up(PICO_DEFAULT_I2C_SDA_PIN);
    gpio_pull_up(PICO_DEFAULT_I2C_SCL_PIN);

    // Make the I2C pins available to picotool
    bi_decl(bi_2pins_with_func(PICO_DEFAULT_I2C_SDA_PIN, PICO_DEFAULT_I2C_SCL_PIN, GPIO_FUNC_I2C));

    printf("Hello, PA1010D! Reading raw data from module...\n");

    pa1010d_write_command(init_command, sizeof(init_command));

    while (1) {
        // Clear array
        memset(numcommand, 0, max_read);
        // Read and re-format
        pa1010d_read_raw(numcommand);
        pa1010d_parse_string(numcommand, "GNRMC");

        // Wait for data to refresh
        sleep_ms(1000);

        // Clear terminal
        printf("\e[1;1H\e[2J");
    }
#endif
    return 0;
}
```

Bill of Materials

Table 23. A list of materials required for the example

Item	Quantity	Details
Breadboard	1	generic part
Raspberry Pi Pico	1	https://www.raspberrypi.com/products/raspberry-pi-pico/
PA1010D board	1	https://shop.pimoroni.com/products/pa1010d-gps-breakout
M/M Jumper wires	4	generic part

Attaching a PCF8523 Real Time Clock via I2C

This example code shows how to interface the Raspberry Pi Pico to the PCF8523 Real Time Clock

This example allows you to initialise the current time and date and then displays it every half-second. Additionally it lets you set an alarm for a particular time and date and raises an alert accordingly. More information about the module is available at https://learn.adafruit.com/adafruit-pcf8523-real-time-clock.

Wiring information

Wiring up the device requires 4 jumpers, to connect VDD, GND, SDA and SCL. The example here uses I2C port 0, which is assigned to GPIO 4 (SDA) and 5 (SCL) in software. Power is supplied from the 5V pin.

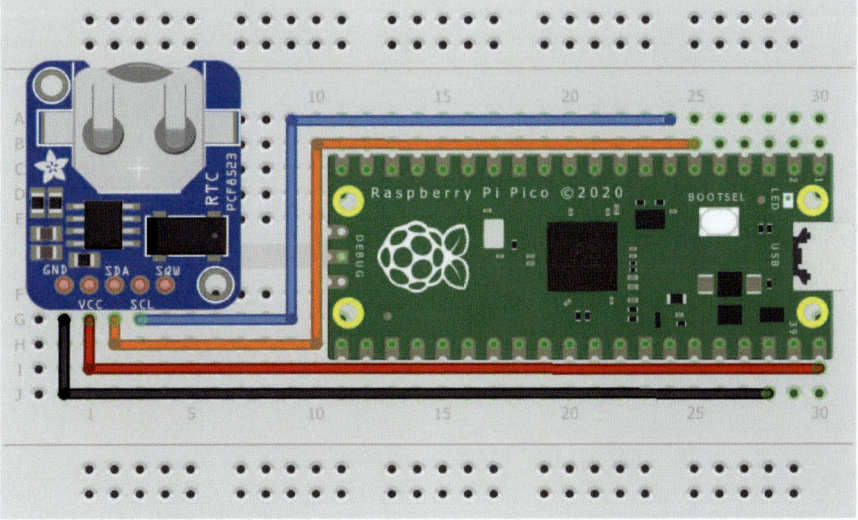

Figure 25. Wiring Diagram for PCF8523.

List of Files

CMakeLists.txt

   CMake file to incorporate the example in to the examples build tree.

Pico Examples: https://github.com/raspberrypi/pico-examples/blob/master/i2c/pcf8523_i2c/CMakeLists.txt

```
1  add_executable(pcf8523_i2c
2          pcf8523_i2c.c
```

```cmake
  3          )
  4
  5  # pull in common dependencies and additional i2c hardware support
  6  target_link_libraries(pcf8523_i2c pico_stdlib hardware_i2c)
  7
  8  # create map/bin/hex file etc.
  9  pico_add_extra_outputs(pcf8523_i2c)
 10
 11  # add url via pico_set_program_url
 12  example_auto_set_url(pcf8523_i2c)
```

pcf8523_i2c.c

The example code.

Pico Examples: *https://github.com/raspberrypi/pico-examples/blob/master/i2c/pcf8523_i2c/pcf8523_i2c.c*

```c
 1  /**
 2   * Copyright (c) 2020 Raspberry Pi (Trading) Ltd.
 3   *
 4   * SPDX-License-Identifier: BSD-3-Clause
 5   */
 6
 7  #include <stdio.h>
 8  #include <string.h>
 9  #include "pico/stdlib.h"
10  #include "pico/binary_info.h"
11  #include "hardware/i2c.h"
12
13  /* Example code to talk to a PCF8520 Real Time Clock module
14
15     Connections on Raspberry Pi Pico board, other boards may vary.
16
17     GPIO PICO_DEFAULT_I2C_SDA_PIN (On Pico this is 4 (physical pin 6)) -> SDA on PCF8520 board
18     GPIO PICO_DEFAULT_I2C_SCK_PIN (On Pico this is 5 (physical pin 7)) -> SCL on PCF8520 board
19     5V (physical pin 40) -> VCC on PCF8520 board
20     GND (physical pin 38)  -> GND on PCF8520 board
21  */
22
23  #ifdef i2c_default
24
25  // By default these devices  are on bus address 0x68
26  static int addr = 0x68;
27
28  static void pcf8520_reset() {
29      // Two byte reset. First byte register, second byte data
30      // There are a load more options to set up the device in different ways that could be added here
31      uint8_t buf[] = {0x00, 0x58};
32      i2c_write_blocking(i2c_default, addr, buf, 2, false);
33  }
34
35  static void pcf820_write_current_time() {
36      // buf[0] is the register to write to
37      // buf[1] is the value that will be written to the register
38      uint8_t buf[2];
39
40      //Write values for the current time in the array
41      //index 0 -> second: bits 4-6 are responsible for the ten's digit and bits 0-3 for the unit's digit
42      //index 1 -> minute: bits 4-6 are responsible for the ten's digit and bits 0-3 for the unit's digit
```

```c
43      //index 2 -> hour: bits 4-5 are responsible for the ten's digit and bits 0-3 for the unit's digit
44      //index 3 -> day of the month: bits 4-5 are responsible for the ten's digit and bits 0-3 for the unit's digit
45      //index 4 -> day of the week: where Sunday = 0x00, Monday = 0x01, Tuesday... ...Saturday = 0x06
46      //index 5 -> month: bit 4 is responsible for the ten's digit and bits 0-3 for the unit's digit
47      //index 6 -> year: bits 4-7 are responsible for the ten's digit and bits 0-3 for the unit's digit
48
49      //NOTE: if the value in the year register is a multiple for 4, it will be considered a leap year and hence will include the 29th of February
50
51      uint8_t current_val[7] = {0x00, 0x00, 0x00, 0x00, 0x00, 0x00, 0x00};
52
53      for (int i = 3; i < 10; ++i) {
54          buf[0] = i;
55          buf[1] = current_val[i - 3];
56          i2c_write_blocking(i2c_default, addr, buf, 2, false);
57      }
58 }
59
60 static void pcf8520_read_raw(uint8_t *buffer) {
61     // For this particular device, we send the device the register we want to read
62     // first, then subsequently read from the device. The register is auto incrementing
63     // so we don't need to keep sending the register we want, just the first.
64
65     // Start reading acceleration registers from register 0x3B for 6 bytes
66     uint8_t val = 0x03;
67     i2c_write_blocking(i2c_default, addr, &val, 1, true); // true to keep master control of bus
68     i2c_read_blocking(i2c_default, addr, buffer, 7, false);
69 }
70
71
72 void pcf8520_set_alarm() {
73     // buf[0] is the register to write to
74     // buf[1] is the value that will be written to the register
75     uint8_t buf[2];
76
77     // Default value of alarm register is 0x80
78     // Set bit 8 of values to 0 to activate that particular alarm
79     // Index 0 -> minute: bits 4-5 are responsible for the ten's digit and bits 0-3 for the unit's digit
80     // Index 1 -> hour: bits 4-6 are responsible for the ten's digit and bits 0-3 for the unit's digit
81     // Index 2 -> day of the month: bits 4-5 are responsible for the ten's digit and bits 0-3 for the unit's digit
82     // Index 3 -> day of the week: where Sunday = 0x00, Monday = 0x01, Tuesday... ...Saturday = 0x06
83
84     uint8_t alarm_val[4] = {0x01, 0x80, 0x80, 0x80};
85     // Write alarm values to registers
86     for (int i = 10; i < 14; ++i) {
87         buf[0] = (uint8_t) i;
88         buf[1] = alarm_val[i - 10];
89         i2c_write_blocking(i2c_default, addr, buf, 2, false);
90     }
91 }
92
93 void pcf8520_check_alarm() {
94     // Check bit 3 of control register 2 for alarm flags
```

```c
95      uint8_t status[1];
96      uint8_t val = 0x01;
97      i2c_write_blocking(i2c_default, addr, &val, 1, true); // true to keep master control of
   bus
98      i2c_read_blocking(i2c_default, addr, status, 1, false);
99
100     if ((status[0] & 0x08) == 0x08) {
101         printf("ALARM RINGING");
102     } else {
103         printf("Alarm not triggered yet");
104     }
105 }
106
107
108 void pcf8520_convert_time(int conv_time[7], const uint8_t raw_time[7]) {
109     // Convert raw data into time
110     conv_time[0] = (10 * (int) ((raw_time[0] & 0x70) >> 4)) + ((int) (raw_time[0] & 0x0F));
111     conv_time[1] = (10 * (int) ((raw_time[1] & 0x70) >> 4)) + ((int) (raw_time[1] & 0x0F));
112     conv_time[2] = (10 * (int) ((raw_time[2] & 0x30) >> 4)) + ((int) (raw_time[2] & 0x0F));
113     conv_time[3] = (10 * (int) ((raw_time[3] & 0x30) >> 4)) + ((int) (raw_time[3] & 0x0F));
114     conv_time[4] = (int) (raw_time[4] & 0x07);
115     conv_time[5] = (10 * (int) ((raw_time[5] & 0x10) >> 4)) + ((int) (raw_time[5] & 0x0F));
116     conv_time[6] = (10 * (int) ((raw_time[6] & 0xF0) >> 4)) + ((int) (raw_time[6] & 0x0F));
117 }
118 #endif
119
120 int main() {
121     stdio_init_all();
122 #if !defined(i2c_default) || !defined(PICO_DEFAULT_I2C_SDA_PIN) ||
    !defined(PICO_DEFAULT_I2C_SCL_PIN)
123 #warning i2c/pcf8520_i2c example requires a board with I2C pins
124     puts("Default I2C pins were not defined");
125 #else
126     printf("Hello, PCF8520! Reading raw data from registers...\n");
127
128     // This example will use I2C0 on the default SDA and SCL pins (4, 5 on a Pico)
129     i2c_init(i2c_default, 400 * 1000);
130     gpio_set_function(PICO_DEFAULT_I2C_SDA_PIN, GPIO_FUNC_I2C);
131     gpio_set_function(PICO_DEFAULT_I2C_SCL_PIN, GPIO_FUNC_I2C);
132     gpio_pull_up(PICO_DEFAULT_I2C_SDA_PIN);
133     gpio_pull_up(PICO_DEFAULT_I2C_SCL_PIN);
134     // Make the I2C pins available to picotool
135     bi_decl(bi_2pins_with_func(PICO_DEFAULT_I2C_SDA_PIN, PICO_DEFAULT_I2C_SCL_PIN,
    GPIO_FUNC_I2C));
136
137     pcf8520_reset();
138
139     pcf820_write_current_time();
140     pcf8520_set_alarm();
141     pcf8520_check_alarm();
142
143     uint8_t raw_time[7];
144     int real_time[7];
145     char days_of_week[7][12] = {"Sunday", "Monday", "Tuesday", "Wednesday", "Thursday",
    "Friday", "Saturday"};
146
147     while (1) {
148
149         pcf8520_read_raw(raw_time);
150         pcf8520_convert_time(real_time, raw_time);
151
152         printf("Time: %02d : %02d : %02d\n", real_time[2], real_time[1], real_time[0]);
153         printf("Date: %s %02d / %02d / %02d\n", days_of_week[real_time[4]], real_time[3],
```

```
            real_time[5], real_time[6]);
154         pcf8520_check_alarm();
155
156         sleep_ms(500);
157
158         // Clear terminal
159         printf("\e[1;1H\e[2J");
160     }
161 #endif
162     return 0;
163 }
```

Bill of Materials

Table 24. A list of materials required for the example

Item	Quantity	Details
Breadboard	1	generic part
Raspberry Pi Pico	1	https://www.raspberrypi.com/products/raspberry-pi-pico/
PCF8523 board	1	https://www.adafruit.com/product/3295
M/M Jumper wires	4	generic part

Appendix B: SDK configuration

SDK configuration is the process of customising the SDK code for your particular build/application. As the parts of the SDK that you use are recompiled as part of your build, configuration options can be chosen at compile time resulting in smaller and more efficient customized versions of the code.

This chapter will show what configuration parameters are available, and how they can be changed.

SDK configuration parameters are passed as C preprocessor definitions to the build. The most common way to override them is to specify them in your `CMakeLists.txt` when you define your executable or library:

e.g.

```
add_executable(my_program main.c)
...
target_compile_definitions(my_program PRIVATE
    PICO_STACK_SIZE=4096
)
```

or if you are creating a library, and you want to add compile definitions whenever your library is included:

```
add_library(my_library INTERFACE)
...
target_compile_definitions(my_library INTERFACE
    PICO_STDIO_DEFAULT_CRLF=0
    PICO_DEFAULT_UART=1
)
```

The definitions can also be overridden in header files, as is commonly done for board configuration (see Appendix D).

For example,. the Pimoroni Tiny2040 board header configures the following to specify appropriate board settings for the default I2C channel exposed on that board.

```
// --- I2C ---
#ifndef PICO_DEFAULT_I2C
#define PICO_DEFAULT_I2C 1
#endif
#ifndef PICO_DEFAULT_I2C_SDA_PIN
#define PICO_DEFAULT_I2C_SDA_PIN 2
#endif
#ifndef PICO_DEFAULT_I2C_SCL_PIN
#define PICO_DEFAULT_I2C_SCL_PIN 3
#endif
```

> **NOTE**
>
> The `#ifdef` allows these values to still be overridden by the build (i.e. in `CMakeLists.txt`)

If you would rather set values in your own header file rather than via `CMake`, then you must make sure the header is included by all compilation (including the SDK sources). Using a custom `PICO_BOARD` header is one way of doing this, but a more advanced way is to have the SDK include your header via `pico/config.h` which itself is included by every SDK source file.

This can be done by adding the following before the `pico_sdk_init()` in your `CMakeLists.txt`:

```
list(APPEND PICO_CONFIG_HEADER_FILES path/to/your/header.h)
```

Configuration Parameters

Table 25. SDK and Board Configuration Parameters

Parameter name	Defined in	Default	Description
CYW43_ARCH_DEBUG_ENABLED	cyw43_arch.h	1 in debug builds	Enable/disable some debugging output in the pico_cyw43_arch module
GPIO_IRQ_CALLBACK_ORDER_PRIORITY	gpio.h	PICO_SHARED_IRQ_HANDLER_LOWEST_ORDER_PRIORITY	the irq priority order of the default IRQ callback
GPIO_RAW_IRQ_HANDLER_DEFAULT_ORDER_PRIORITY	gpio.h	PICO_SHARED_IRQ_HANDLER_DEFAULT_ORDER_PRIORITY	the irq priority order of raw IRQ handlers if the priortiy is not specified
PARAM_ASSERTIONS_DISABLE_ALL	assert.h	0	Global assert disable
PARAM_ASSERTIONS_ENABLED_ADC	adc.h	0	Enable/disable assertions in the ADC module
PARAM_ASSERTIONS_ENABLED_ADDRESS_ALIAS	address_mapped.h	0	Enable/disable assertions in memory address aliasing macros
PARAM_ASSERTIONS_ENABLED_CLOCKS	clocks.h	0	Enable/disable assertions in the clocks module
PARAM_ASSERTIONS_ENABLED_CYW43_ARCH	cyw43_arch.h	0	Enable/disable assertions in the pico_cyw43_arch module
PARAM_ASSERTIONS_ENABLED_DMA	dma.h	0	Enable/disable DMA assertions
PARAM_ASSERTIONS_ENABLED_EXCEPTION	exception.h	0	Enable/disable assertions in the exception module
PARAM_ASSERTIONS_ENABLED_FLASH	flash.h	0	Enable/disable assertions in the flash module
PARAM_ASSERTIONS_ENABLED_GPIO	gpio.h	0	Enable/disable assertions in the GPIO module
PARAM_ASSERTIONS_ENABLED_I2C	i2c.h	0	Enable/disable assertions in the I2C module
PARAM_ASSERTIONS_ENABLED_INTERP	interp.h	0	Enable/disable assertions in the interpolation module
PARAM_ASSERTIONS_ENABLED_IRQ	irq.h	0	Enable/disable assertions in the IRQ module
PARAM_ASSERTIONS_ENABLED_LOCK_CORE	lock_core.h	0	Enable/disable assertions in the lock core

Parameter name	Defined in	Default	Description
PARAM_ASSERTIONS_ENABLED_PHEAP	pheap.h	0	Enable/disable assertions in the pheap module
PARAM_ASSERTIONS_ENABLED_PIO	pio.h	0	Enable/disable assertions in the PIO module
PARAM_ASSERTIONS_ENABLED_PIO_INSTRUCTIONS	pio_instructions.h	0	Enable/disable assertions in the PIO instructions
PARAM_ASSERTIONS_ENABLED_PWM	pwm.h	0	Enable/disable assertions in the PWM module
PARAM_ASSERTIONS_ENABLED_SPI	spi.h	0	Enable/disable assertions in the SPI module
PARAM_ASSERTIONS_ENABLED_SYNC	sync.h	0	Enable/disable assertions in the HW sync module
PARAM_ASSERTIONS_ENABLED_TIME	time.h	0	Enable/disable assertions in the time module
PARAM_ASSERTIONS_ENABLED_TIMER	timer.h	0	Enable/disable assertions in the timer module
PARAM_ASSERTIONS_ENABLED_UART	uart.h	0	Enable/disable assertions in the UART module
PARAM_ASSERTIONS_ENABLE_ALL	assert.h	0	Global assert enable
PICO_BOOTSEL_VIA_DOUBLE_RESET_ACTIVITY_LED	pico_bootsel_via_double_reset.c		Optionally define a pin to use as bootloader activity LED when BOOTSEL mode is entered via reset double tap
PICO_BOOTSEL_VIA_DOUBLE_RESET_INTERFACE_DISABLE_MASK	pico_bootsel_via_double_reset.c	0	Optionally disable either the mass storage interface (bit 0) or the PICOBOOT interface (bit 1) when entering BOOTSEL mode via double reset
PICO_BOOTSEL_VIA_DOUBLE_RESET_TIMEOUT_MS	pico_bootsel_via_double_reset.c	200	Window of opportunity for a second press of a reset button to enter BOOTSEL mode (milliseconds)
PICO_BOOT_STAGE2_CHOOSE_AT25SF128A	config.h	0	Select boot2_at25sf128a as the boot stage 2 when no boot stage 2 selection is made by the CMake build
PICO_BOOT_STAGE2_CHOOSE_GENERIC_03H	config.h	1	Select boot2_generic_03h as the boot stage 2 when no boot stage 2 selection is made by the CMake build
PICO_BOOT_STAGE2_CHOOSE_IS25LP080	config.h	0	Select boot2_is25lp080 as the boot stage 2 when no boot stage 2 selection is made by the CMake build
PICO_BOOT_STAGE2_CHOOSE_W25Q080	config.h	0	Select boot2_w25q080 as the boot stage 2 when no boot stage 2 selection is made by the CMake build

Parameter name	Defined in	Default	Description
PICO_BOOT_STAGE2_CHOOSE_W25X10CL	config.h	0	Select boot2_w25x10cl as the boot stage 2 when no boot stage 2 selection is made by the CMake build
PICO_BUILD_BOOT_STAGE2_NAME	config.h		The name of the boot stage 2 if selected by the build
PICO_CMSIS_RENAME_EXCEPTIONS	rename_exceptions.h	1	Whether to rename SDK exceptions such as isr_nmi to their CMSIS equivalent i.e. NMI_Handler
PICO_CONFIG_HEADER	pico.h		unquoted path to header include in place of the default pico/config.h which may be desirable for build systems which can't easily generate the config_autogen header
PICO_CONFIG_RTOS_ADAPTER_HEADER	config.h		unquoted path to header include in the default pico/config.h for RTOS integration defines that must be included in all sources
PICO_CORE1_STACK_SIZE	multicore.h	PICO_STACK_SIZE (0x800)	Stack size for core 1
PICO_CYW43_ARCH_DEFAULT_COUNTRY_CODE	cyw43_arch.h	CYW43_COUNTRY_WORLDWIDE	Default country code for the cyw43 wireless driver
PICO_DEBUG_MALLOC	malloc.h	0	Enable/disable debug printf from malloc
PICO_DEBUG_MALLOC_LOW_WATER	malloc.h	0	Define the lower bound for allocation addresses to be printed by PICO_DEBUG_MALLOC
PICO_DEBUG_PIN_BASE	gpio.h	19	First pin to use for debug output (if enabled)
PICO_DEBUG_PIN_COUNT	gpio.h	3	Number of pins to use for debug output (if enabled)
PICO_DEFAULT_I2C	i2c.h		Define the default I2C for a board
PICO_DEFAULT_I2C_SCL_PIN	i2c.h		Define the default I2C SCL pin
PICO_DEFAULT_I2C_SDA_PIN	i2c.h		Define the default I2C SDA pin
PICO_DEFAULT_IRQ_PRIORITY	irq.h	0x80	Define the default IRQ priority
PICO_DEFAULT_LED_PIN	stdlib.h		Optionally define a pin that drives a regular LED on the board
PICO_DEFAULT_LED_PIN_INVERTED	stdlib.h	0	1 if LED is inverted or 0 if not
PICO_DEFAULT_SPI	spi.h		Define the default SPI for a board
PICO_DEFAULT_SPI_CSN_PIN	spi.h		Define the default SPI CSN pin
PICO_DEFAULT_SPI_RX_PIN	spi.h		Define the default SPI RX pin
PICO_DEFAULT_SPI_SCK_PIN	spi.h		Define the default SPI SCK pin
PICO_DEFAULT_SPI_TX_PIN	spi.h		Define the default SPI TX pin

Parameter name	Defined in	Default	Description
PICO_DEFAULT_UART	uart.h		Define the default UART used for printf etc
PICO_DEFAULT_UART_BAUD_RATE	uart.h	115200	Define the default UART baudrate
PICO_DEFAULT_UART_RX_PIN	uart.h		Define the default UART RX pin
PICO_DEFAULT_UART_TX_PIN	uart.h		Define the default UART TX pin
PICO_DEFAULT_WS2812_PIN	stdlib.h		Optionally define a pin that controls data to a WS2812 compatible LED on the board
PICO_DEFAULT_WS2812_POWER_PIN	stdlib.h		Optionally define a pin that controls power to a WS2812 compatible LED on the board
PICO_DISABLE_SHARED_IRQ_HANDLERS	irq.h	0	Disable shared IRQ handlers
PICO_DOUBLE_SUPPORT_ROM_V1	platform.h	1	Include double support code for RP2040 B0 when that chip revision is supported
PICO_FLASH_SIZE_BYTES	flash.h		size of primary flash in bytes
PICO_FLOAT_SUPPORT_ROM_V1	platform.h	1	Include float support code for RP2040 B0 when that chip revision is supported
PICO_HEAP_SIZE	platform.h	0x800	Heap size to reserve
PICO_MALLOC_PANIC	malloc.h	1	Enable/disable panic when an allocation failure occurs
PICO_MAX_SHARED_IRQ_HANDLERS	irq.h	4	Maximum number of shared IRQ handlers
PICO_NO_FPGA_CHECK	platform.h	0	Remove the FPGA platform check for small code size reduction
PICO_NO_RAM_VECTOR_TABLE	platform.h	0	Enable/disable the RAM vector table
PICO_PANIC_FUNCTION	runtime.c		Name of a function to use in place of the stock panic function or empty string to simply breakpoint on panic
PICO_PHEAP_MAX_ENTRIES	pheap.h	255	Maximum number of entries in the pheap
PICO_PRINTF_ALWAYS_INCLUDED	printf.h	1 in debug build 0 otherwise	Whether to always include printf code even if only called weakly (by panic)
PICO_PRINTF_DEFAULT_FLOAT_PRECISION	printf.c	6	Define default floating point precision
PICO_PRINTF_FTOA_BUFFER_SIZE	printf.c	32	Define printf ftoa buffer size
PICO_PRINTF_MAX_FLOAT	printf.c	1e9	Define the largest float suitable to print with %f
PICO_PRINTF_NTOA_BUFFER_SIZE	printf.c	32	Define printf ntoa buffer size
PICO_PRINTF_SUPPORT_EXPONENTIAL	printf.c	1	Enable exponential floating point printing

Parameter name	Defined in	Default	Description
PICO_PRINTF_SUPPORT_FLOAT	printf.c	1	Enable floating point printing
PICO_PRINTF_SUPPORT_LONG_LONG	printf.c	1	Enable support for long long types (%llu or %p)
PICO_PRINTF_SUPPORT_PTRDIFF_T	printf.c	1	Enable support for the ptrdiff_t type (%t)
PICO_QUEUE_MAX_LEVEL	queue.h	0	Maintain a field for the highest level that has been reached by a queue
PICO_RP2040_B0_SUPPORTED	platform.h	1	Whether to include any specific software support for RP2040 B0 revision
PICO_RP2040_B1_SUPPORTED	platform.h	1	Whether to include any specific software support for RP2040 B1 revision
PICO_RP2040_B2_SUPPORTED	platform.h	1	Whether to include any specific software support for RP2040 B2 revision
PICO_SHARED_IRQ_HANDLER_DEFAULT_ORDER_PRIORITY	irq.h	0x80	Set default shared IRQ order priority
PICO_SPINLOCK_ID_CLAIM_FREE_FIRST	sync.h	24	Lowest Spinlock ID in the 'claim free' range
PICO_SPINLOCK_ID_CLAIM_FREE_LAST	sync.h	31	Highest Spinlock ID in the 'claim free' range
PICO_SPINLOCK_ID_HARDWARE_CLAIM	sync.h	11	Spinlock ID for Hardware claim protection
PICO_SPINLOCK_ID_IRQ	sync.h	9	Spinlock ID for IRQ protection
PICO_SPINLOCK_ID_OS1	sync.h	14	First Spinlock ID reserved for use by low level OS style software
PICO_SPINLOCK_ID_OS2	sync.h	15	Second Spinlock ID reserved for use by low level OS style software
PICO_SPINLOCK_ID_STRIPED_FIRST	sync.h	16	Lowest Spinlock ID in the 'striped' range
PICO_SPINLOCK_ID_STRIPED_LAST	sync.h	23	Highest Spinlock ID in the 'striped' range
PICO_SPINLOCK_ID_TIMER	sync.h	10	Spinlock ID for Timer protection
PICO_STACK_SIZE	platform.h	0x800	Stack Size
PICO_STDIO_DEFAULT_CRLF	stdio.h	1	Default for CR/LF conversion enabled on all stdio outputs
PICO_STDIO_ENABLE_CRLF_SUPPORT	stdio.h	1	Enable/disable CR/LF output conversion support
PICO_STDIO_SEMIHOSTING_DEFAULT_CRLF	stdio_semihosting.h	PICO_STDIO_DEFAULT_CRLF	Default state of CR/LF translation for semihosting output

Parameter name	Defined in	Default	Description
PICO_STDIO_STACK_BUFFER_SIZE	stdio.h	128	Define printf buffer size (on stack)… this is just a working buffer not a max output size
PICO_STDIO_UART_DEFAULT_CRLF	stdio_uart.h	PICO_STDIO_DEFAULT_CRLF	Default state of CR/LF translation for UART output
PICO_STDIO_USB_CONNECT_WAIT_TIMEOUT_MS	stdio_usb.h	0	Maximum number of milliseconds to wait during initialization for a CDC connection from the host (negative means indefinite) during initialization
PICO_STDIO_USB_DEFAULT_CRLF	stdio_usb.h	PICO_STDIO_DEFAULT_CRLF	Default state of CR/LF translation for USB output
PICO_STDIO_USB_ENABLE_RESET_VIA_BAUD_RATE	stdio_usb.h	1	Enable/disable resetting into BOOTSEL mode if the host sets the baud rate to a magic value (PICO_STDIO_USB_RESET_MAGIC_BAUD_RATE)
PICO_STDIO_USB_ENABLE_RESET_VIA_VENDOR_INTERFACE	stdio_usb.h	1	Enable/disable resetting into BOOTSEL mode via an additional VENDOR USB interface - enables picotool based reset
PICO_STDIO_USB_LOW_PRIORITY_IRQ	stdio_usb.h		Explicit User IRQ number to claim for tud_task() background execution instead of letting the implementation pick a free one dynamically (deprecated)
PICO_STDIO_USB_POST_CONNECT_WAIT_DELAY_MS	stdio_usb.h	50	Number of extra milliseconds to wait when using PICO_STDIO_USB_CONNECT_WAIT_TIMEOUT_MS after a host CDC connection is detected (some host terminals seem to sometimes lose transmissions sent right after connection)
PICO_STDIO_USB_RESET_BOOTSEL_ACTIVITY_LED	stdio_usb.h		Optionally define a pin to use as bootloader activity LED when BOOTSEL mode is entered via USB (either VIA_BAUD_RATE or VIA_VENDOR_INTERFACE)
PICO_STDIO_USB_RESET_BOOTSEL_FIXED_ACTIVITY_LED	stdio_usb.h	0	Whether the pin specified by PICO_STDIO_USB_RESET_BOOTSEL_ACTIVITY_LED is fixed or can be modified by picotool over the VENDOR USB interface
PICO_STDIO_USB_RESET_BOOTSEL_INTERFACE_DISABLE_MASK	stdio_usb.h	0	Optionally disable either the mass storage interface (bit 0) or the PICOBOOT interface (bit 1) when entering BOOTSEL mode via USB (either VIA_BAUD_RATE or VIA_VENDOR_INTERFACE)

Parameter name	Defined in	Default	Description
PICO_STDIO_USB_RESET_INTERFACE_SUPPORT_RESET_TO_BOOTSEL	stdio_usb.h	1	If vendor reset interface is included allow rebooting to BOOTSEL mode
PICO_STDIO_USB_RESET_INTERFACE_SUPPORT_RESET_TO_FLASH_BOOT	stdio_usb.h	1	If vendor reset interface is included allow rebooting with regular flash boot
PICO_STDIO_USB_RESET_MAGIC_BAUD_RATE	stdio_usb.h	1200	baud rate that if selected causes a reset into BOOTSEL mode (if PICO_STDIO_USB_ENABLE_RESET_VIA_BAUD_RATE is set)
PICO_STDIO_USB_RESET_RESET_TO_FLASH_DELAY_MS	stdio_usb.h	100	delays in ms before rebooting via regular flash boot
PICO_STDIO_USB_STDOUT_TIMEOUT_US	stdio_usb.h	500000	Number of microseconds to be blocked trying to write USB output before assuming the host has disappeared and discarding data
PICO_STDIO_USB_TASK_INTERVAL_US	stdio_usb.h	1000	Period of microseconds between calling tud_task in the background
PICO_STDOUT_MUTEX	stdio.h	1	Enable/disable mutex around stdout
PICO_TIME_DEFAULT_ALARM_POOL_DISABLED	time.h	0	Disable the default alarm pool
PICO_TIME_DEFAULT_ALARM_POOL_HARDWARE_ALARM_NUM	time.h	3	Select which HW alarm is used for the default alarm pool
PICO_TIME_DEFAULT_ALARM_POOL_MAX_TIMERS	time.h	16	Selects the maximum number of concurrent timers in the default alarm pool
PICO_TIME_SLEEP_OVERHEAD_ADJUST_US	time.h	6	How many microseconds to wake up early (and then busy_wait) to account for timer overhead when sleeping in low power mode
PICO_UART_DEFAULT_CRLF	uart.h	0	Enable/disable CR/LF translation on UART
PICO_UART_ENABLE_CRLF_SUPPORT	uart.h	1	Enable/disable CR/LF translation support
PICO_USE_MALLOC_MUTEX	malloc.h	1 with pico_multicore, 0 otherwise	Whether to protect malloc etc with a mutex
PICO_XOSC_STARTUP_DELAY_MULTIPLIER	xosc.h	1	Multiplier to lengthen xosc startup delay to accommodate slow-starting oscillators
USB_DPRAM_MAX	usb.h	4096	Set amount of USB RAM used by USB system
XOSC_MHZ	platform_defs.h	12	The crystal oscillator frequency in Mhz

Appendix C: CMake build configuration

CMake configuration variables can be used to customize the way the SDK performs builds

Configuration Parameters

Table 26. CMake Configuration Variables

Parameter name	Defined in	Default	Description
PICO_BARE_METAL	CMakeLists.txt	0	Flag to exclude anything except base headers from the build
PICO_BOARD	board_setup.cmake	pico	The board name being built for. This is overridable from the user environment
PICO_BOARD_CMAKE_DIRS	board_setup.cmake	""	Directories to look for <PICO_BOARD>.cmake in. This is overridable from the user environment
PICO_BOARD_HEADER_DIRS	generic_board.cmake	""	Directories to look for <PICO_BOARD>.h in. This is overridable from the user environment
PICO_CMAKE_RELOAD_PLATFORM_FILE	pico_pre_load_platform.cmake	none	custom CMake file to use to set up the platform environment
PICO_COMPILER	pico_pre_load_toolchain.cmake	none	Optionally specifies a different compiler (other than pico_arm_gcc.cmake) - this is not yet fully supported
PICO_CONFIG_HEADER_FILES	CMakeLists.txt	""	List of extra header files to include from pico/config.h for all platforms
PICO_CONFIG_HOST_HEADER_FILES	CMakeLists.txt	""	List of extra header files to include from pico/config.h for host platform
PICO_CONFIG_RP2040_HEADER_FILES	CMakeLists.txt	""	List of extra header files to include from pico/config.h for rp2040 platform
PICO_CXX_ENABLE_CXA_ATEXIT	CMakeLists.txt	0	Enabled cxa-atexit
PICO_CXX_ENABLE_EXCEPTIONS	CMakeLists.txt	0	Enabled CXX exception handling
PICO_CXX_ENABLE_RTTI	CMakeLists.txt	0	Enabled CXX rtti
PICO_DEFAULT_BOOT_STAGE2_FILE	CMakeLists.txt	…/boot2_w25q080.S	Default stage2 file to use unless overridden by pico_set_boot_stage2 on the TARGET
PICO_NO_GC_SECTIONS	CMakeLists.txt	0	Disable -ffunction-sections -fdata-sections and --gc-sections
PICO_NO_HARDWARE	rp2_common.cmake	1 for PICO_PLATFORM host 0 otherwise	OPTION: Whether the build is not targeting an RP2040 device

Parameter name	Defined in	Default	Description
PICO_NO_TARGET_NAME	rp2_common.cmake	0	Don't defined PICO_TARGET_NAME
PICO_NO_UF2	rp2_common.cmake	0	Disable UF2 output
PICO_ON_DEVICE	rp2_common.cmake	0 for PICO_PLATFORM host 1 otherwise	OPTION: Whether the build is targeting an RP2040 device
PICO_PLATFORM	pico_pre_load_platform.cmake	rp2040 or environment value	platform to build for e.g. rp2040/host
PICO_STDIO_SEMIHOSTING	CMakeLists.txt	0	OPTION: Globally enable stdio semihosting
PICO_STDIO_UART	CMakeLists.txt	1	OPTION: Globally enable stdio UART
PICO_STDIO_USB	CMakeLists.txt	0	OPTION: Globally enable stdio USB
PICO_TOOLCHAIN_PATH	pico_pre_load_toolchain.cmake	none (i.e. search system paths)	Path to search for compiler

Control of binary type produced (advanced)

These variables control how executables for RP2040 are laid out in memory. The default is for the code and data to be entirely stored in flash with writable data (and some specifically marked) methods to copied into RAM at startup.

PICO_DEFAULT_BINARY_TYPE	default	The default is flash binaries which are stored in and run from flash.
	no_flash	This option selects a RAM only binaries, that does not require any flash. Note: this type of binary must be loaded on each device reboot via a UF2 file or from the debugger.
	copy_to_ram	This option selects binaries which are stored in flash, but copy themselves to RAM before executing.
	blocked_ram	
PICO_NO_FLASH*	0 / 1	Equivalent to PICO_DEFAULT_BINARY_TYPE=no_flash if 1
PICO_COPY_TO_RAM*	0 / 1	Equivalent to PICO_DEFAULT_BINARY_TYPE=copy_to_ram if 1
PICO_USE_BLOCKED_RAM*	0 / 1	Equivalent to PICO_DEFAULT_BINARY_TYPE=blocked_ram if 1

 TIP

The binary type can be set on a per executable target (as created by `add_executable`) basis by calling `pico_set_binary_type(target type)` where type is the same as for `PICO_DEFAULT_BINARY_TYPE`

Appendix D: Board configuration

Board Configuration

Board configuration is the process of customising the SDK to run on a specific board design. The SDK comes with some predefined configurations for boards produced by Raspberry Pi and other manufacturers, the main (and default) example being the Raspberry Pi Pico.

Configurations specify a number of parameters that could vary between hardware designs. For example, default UART ports, on-board LED locations and flash capacities etc.

This chapter will go through where these configurations files are, how to make changes and set parameters, and how to build your SDK using CMake with your customisations.

The Configuration files

Board specific configuration files are stored in the SDK source tree, at `…/src/boards/include/boards/<boardname>.h`. The default configuration file is that for the Raspberry Pi Pico, and at the time of writing is:

`<sdk_path>/src/boards/include/boards/pico.h`

This relatively short file contains overrides from default of a small number of parameters used by the SDK when building code.

SDK: https://github.com/raspberrypi/pico-sdk/blob/master/src/boards/include/boards/pico.h

```
1  /*
2   * Copyright (c) 2020 Raspberry Pi (Trading) Ltd.
3   *
4   * SPDX-License-Identifier: BSD-3-Clause
5   */
6
7  // -------------------------------------------------
8  // NOTE: THIS HEADER IS ALSO INCLUDED BY ASSEMBLER SO
9  //       SHOULD ONLY CONSIST OF PREPROCESSOR DIRECTIVES
10 // -------------------------------------------------
11
12 // This header may be included by other board headers as "boards/pico.h"
13
14 #ifndef _BOARDS_PICO_H
15 #define _BOARDS_PICO_H
16
17 // For board detection
18 #define RASPBERRYPI_PICO
19
20 // --- UART ---
21 #ifndef PICO_DEFAULT_UART
22 #define PICO_DEFAULT_UART 0
23 #endif
24 #ifndef PICO_DEFAULT_UART_TX_PIN
25 #define PICO_DEFAULT_UART_TX_PIN 0
26 #endif
27 #ifndef PICO_DEFAULT_UART_RX_PIN
28 #define PICO_DEFAULT_UART_RX_PIN 1
29 #endif
30
31 // --- LED ---
```

```
32  #ifndef PICO_DEFAULT_LED_PIN
33  #define PICO_DEFAULT_LED_PIN 25
34  #endif
35  // no PICO_DEFAULT_WS2812_PIN
36
37  // --- I2C ---
38  #ifndef PICO_DEFAULT_I2C
39  #define PICO_DEFAULT_I2C 0
40  #endif
41  #ifndef PICO_DEFAULT_I2C_SDA_PIN
42  #define PICO_DEFAULT_I2C_SDA_PIN 4
43  #endif
44  #ifndef PICO_DEFAULT_I2C_SCL_PIN
45  #define PICO_DEFAULT_I2C_SCL_PIN 5
46  #endif
47
48  // --- SPI ---
49  #ifndef PICO_DEFAULT_SPI
50  #define PICO_DEFAULT_SPI 0
51  #endif
52  #ifndef PICO_DEFAULT_SPI_SCK_PIN
53  #define PICO_DEFAULT_SPI_SCK_PIN 18
54  #endif
55  #ifndef PICO_DEFAULT_SPI_TX_PIN
56  #define PICO_DEFAULT_SPI_TX_PIN 19
57  #endif
58  #ifndef PICO_DEFAULT_SPI_RX_PIN
59  #define PICO_DEFAULT_SPI_RX_PIN 16
60  #endif
61  #ifndef PICO_DEFAULT_SPI_CSN_PIN
62  #define PICO_DEFAULT_SPI_CSN_PIN 17
63  #endif
64
65  // --- FLASH ---
66
67  #define PICO_BOOT_STAGE2_CHOOSE_W25Q080 1
68
69  #ifndef PICO_FLASH_SPI_CLKDIV
70  #define PICO_FLASH_SPI_CLKDIV 2
71  #endif
72
73  #ifndef PICO_FLASH_SIZE_BYTES
74  #define PICO_FLASH_SIZE_BYTES (2 * 1024 * 1024)
75  #endif
76
77  // Drive high to force power supply into PWM mode (lower ripple on 3V3 at light loads)
78  #define PICO_SMPS_MODE_PIN 23
79
80  #ifndef PICO_RP2040_B0_SUPPORTED
81  #define PICO_RP2040_B0_SUPPORTED 1
82  #endif
83
84  #endif
```

As can be seen, it sets up the default UART to `uart0`, the GPIO pins to be used for that UART, the GPIO pin used for the on-board LED, and the flash size.

To create your own configuration file, create a file in the board `../source/folder` with the name of your board, for example, `myboard.h`. Enter your board specific parameters in this file.

Building applications with a custom board configuration

The CMake system is what specifies which board configuration is going to be used.

To create a new build based on a new board configuration (we will use the `myboard` example from the previous section) first create a new build folder under your project folder. For our example we will use the pico-examples folder.

```
$ cd pico-examples
$ mkdir myboard_build
$ cd myboard_build
```

then run cmake as follows:

```
$ cmake -D"PICO_BOARD=myboard" ..
```

This will set up the system ready to build so you can simply type `make` in the `myboard_build` folder and the examples will be built for your new board configuration.

Available configuration parameters

Table 25 lists all the available configuration parameters available within the SDK. You can set any configuration variable in a board configuration header file, however the convention is to limit that to configuration items directly affected by the board design (e.g. pins, clock frequencies etc.). Other configuration items should generally be overridden in the CMake configuration (or another configuration header) for the application being built.

Appendix E: Building the SDK API documentation

The SDK documentation can be viewed online, but is also part of the SDK itself and can be built directly from the command line. If you haven't already checked out the SDK repository you should do so,

```
$ cd ~/
$ mkdir pico
$ cd pico
$ git clone https://github.com/raspberrypi/pico-sdk.git --branch master
$ cd pico-sdk
$ git submodule update --init
$ cd ..
$ git clone https://github.com/raspberrypi/pico-examples.git --branch master
```

Install doxygen if you don't already have it,

```
$ sudo apt install doxygen
```

Then afterwards you can go ahead and build the documentation,

```
$ cd pico-sdk
$ mkdir build
$ cd build
$ cmake -DPICO_EXAMPLES_PATH=../../pico-examples ..
$ make docs
```

The API documentation will be built and can be found in the `pico-sdk/build/docs/doxygen/html` directory, see Figure 26.

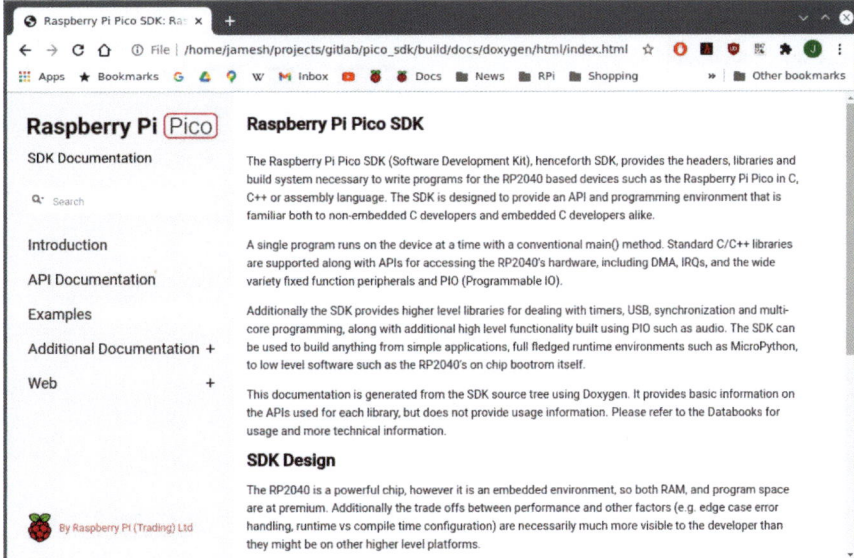

Figure 26. The SDK API documentation

Appendix F: SDK release history

Release 1.0.0 (20/Jan/2021)

Initial release

Release 1.0.1 (01/Feb/2021)

- add `pico_get_unique_id` method to return a unique identifier for a Pico board using the identifier of the external flash
- exposed all 4 pacing timers on the DMA peripheral (previously only 2 were exposed)
- fixed ninja build (i.e. `cmake -G ninja .. ; ninja`)
- minor other improvements and bug fixes

Boot Stage 2

Additionally, a low level change was made to the way flash binaries start executing after `boot_stage2`. This was at the request of folks implementing other language runtimes. It is not generally of concern to end users, however it did require a change to the linker scripts so if you have cloned those to make modifications then you need to port across the relevant changes. If you are porting a different language runtime using the SDK boot_stage2 implementations then you should be aware that you should now have a vector table (rather than executable code) - at `0x10000100`.

Release 1.1.0 (05/Mar/2021)

- Added board headers for Adafruit, Pimoroni & SparkFun boards
    - new values for `PICO_BOARD` are `adafruit_feather_rp2040`, `adafruit_itsybitsy_rp2040`, `adafruit_qtpy_rp2040`, `pimoroni_keybow2040`, `pimoroni_picosystem`, `pimoroni_tiny2040`, `sparkfun_micromod`, `sparkfun_promicro`, `sparkfun_thingplus`, in addition to the existing `pico` and `vgaboard`.
    - Added additional definitions for a default SPI, I2C pins as well as the existing ones for UART
    - Allow *default* pins to be undefined (not all boards have UART for example), and SDK will compile but warn as needed in the absence of default.
    - Added additional definition for a default WS2812 compatible pin (currently unused).
- New reset options
    - Added `pico_bootsel_via_double_reset` library to allow reset to `BOOTSEL` mode via double press of a `RESET` button
    - When using `pico_stdio_usb` i.e. `stdio` connected via USB CDC to host, setting baud rate to 1200 (by default) can optionally be used to reset into `BOOTSEL` mode.
    - When using `pico-stdio_usb` i.e. `stdio` connected via USB CDC to host, an additional interface may be added to give `picotool` control over resetting the device.
- Build improvement for non-SDK or existing library builds
    - Removed additional compiler warnings (register headers now use `_u(x)` macro for unsigned values though).
    - Made build more clang friendly.

This release also contains many bug fixes, documentation updates and minor improvements.

Backwards incompatibility

There are some nominally backwards incompatible changes not worthy of a major version bump:

- `PICO_DEFAULT_UART_` defines now default to undefined if there is no default rather than `-1` previously
- The broken `multicore_sleep_core1()` API has been removed; `multicore_reset_core1` is already available to put core 1 into a deep sleep.

Release 1.1.1 (01/Apr/2021)

This fixes a number of bugs, and additionally adds support for a board configuration header to choose the `boot_stage2`

Release 1.1.2 (07/Apr/2021)

Fixes issues with `boot_stage2` selection

Release 1.2.0 (03/Jun/2021)

This release contains numerous bug fixes and documentation improvements. Additionally it contains the following improvements/notable changes:

> ⚠ **CAUTION**
>
> The `lib/tinyusb` submodule has been updated from 0.8.0 and now tracks upstream https://github.com/hathach/tinyusb.git. It is worth making sure you do a
>
> ```
> git submodule sync
> git submodule update
> ```
>
> to make sure you are correctly tracking upstream TinyUSB if you are not checking out a clean `pico-sdk` repository.
>
> Moving from TinyUSB 0.8.0 to TinyUSB 0.10.1 may require some minor changes to your USB code.

New/improved Board headers

- New board headers support for PICO_BOARDs `arduino_nano_rp240_connect`, `pimoroni_picolipo_4mb` and `pimoroni_picolipo_16mb`
- Missing/new `#defines` for default SPI and I2C pins have been added

Updated TinyUSB to 0.10.1

The `lib/tinyusb` submodule has been updated from 0.8.0 and now tracks upstream https://github.com/hathach/tinyusb.git

Added CMSIS core headers

CMSIS core headers (e.g. `core_cm0plus.h` and `RP2040.h`) are made available via `cmsis_core` INTERFACE library. Additionally, CMSIS standard exception naming is available via `PICO_CMSIS_RENAME_EXCEPTIONS=1`

API improvements

pico_sync

- Added support for recursive mutexes via `recursive_mutex_init()` and `auto_init_recursive_mutex()`
- Added `mutex_enter_timeout_us()`
- Added `critical_section_deinit()`
- Added `sem_acquire_timeout_ms()` and `sem_acquire_block_until()`

hardware_adc

- Added `adc_get_selected_input()`

hardware_clocks

- `clock_get_hz()` now returns actual achieved frequency rather than desired frequency

hardware_dma

- Added `dma_channel_is_claimed()`
- Added new methods for configuring/acknowledging DMA IRQs. `dma_irqn_set_channel_enabled()`, `dma_irqn_set_channel_mask_enabled()`, `dma_irqn_get_channel_status()`, `dma_irqn_acknowledge_channel()` etc.

hardware_exception

New library for setting ARM exception handlers:

- Added `exception_set_exclusive_handler()`, `exception_restore_handler()`, `exception_get_vtable_handler()`

hardware_flash

- Exposed previously private function `flash_do_cmd()` for low level flash command execution

hardware_gpio

- Added `gpio_set_input_hysteresis_enabled()`, `gpio_is_input_hysteresis_enabled()`, `gpio_set_slew_rate()`, `gpio_get_slew_rate()`, `gpio_set_drive_strength()`, `gpio_get_drive_strength()`, `gpio_get_out_level()`, `gpio_set_irqover()`

hardware_i2c

- Corrected a number of incorrect hardware register definitions
- A number of edge cases in the i2c code fixed

hardware_interp

- Added `interp_lane_is_claimed()`, `interp_unclaim_lane_mask()`

hardware_irq

- Notably fixed the `PICO_LOWEST/HIGHEST_IRQ_PRIORITY` values which were backwards!

hardware_pio

- Added new methods for configuring/acknowledging PIO interrupts (`pio_set_irqn_source_enabled()`, `pio_set_irqn_source_mask_enabled()`, `pio_interrupt_get()`, `pio_interrupt_clear()` etc.)
- Added `pio_sm_is_claimed()`

hardware_spi

- Added `spi_get_baudrate()`
- Changed `spi_init()` to return the set/achieved baud rate rather than void
- Changed `spi_is_writable()` to return bool not size_t (it was always 1/0)

hardware_sync

- Notable documentation improvements for spin lock functions
- Added `spin_lock_is_claimed()`

hardware_timer

- Added `busy_wait_ms()` to match `busy_wait_us()`
- Added `hardware_alarm_is_claimed()`

pico_float/pico_double

- Correctly save/restore divider state if floating point is used from interrupts

pico_int64_ops

- Added `PICO_INT64_OPS_IN_RAM` flag to move code into RAM to avoid veneers when calling code is in RAM

pico_runtime

- Added ability to override panic function by setting `PICO_PANIC_FUNCTION=foo` to then use `foo` as the implementation, or setting `PICO_PANIC_FUNCITON=` to simply breakpoint, saving some code space

pico_unique_id

- Added `pico_get_unique_board_id_string()`.

General code improvements

- Removed additional classes of compiler warnings
- Added some missing `const` to method parameters

SVD

- USB DPRAM for device mode is now included

pioasm

- Added `#pragma once` to C/C++ output

RTOS interoperability

Improvements designed to make porting RTOSes either based on the SDK or supporting SDK code easier.

- Added `PICO_DIVIDER_DISABLE_INTERRUPTS` flag to optionally configure all uses of the hardware divider to be guarded by disabling interrupts, rather than requiring on the RTOS to save/restore the divider state on context switch
- Added new abstractions to `pico/lock_core.h` to allow an RTOS to inject replacement code for SDK based low level wait, notify and sleep/timeouts used by synchronization primitives in `pico_sync` and for `sleep_` methods. If an RTOS implements these few simple methods, then all SDK semaphore, mutex, queue, sleep methods can be safely used both within/to/from RTOS tasks, but also to communicate with non-RTOS task aware code, whether it be existing libraries and IRQ handlers or code running perhaps (though not necessarily) on the other core

CMake build changes

Substantive changes have been made to the CMake build, so if you are using a hand crafted non-CMake build, you **will** need to update your compile/link flags. Additionally changed some possibly confusing status messages from CMake build generation to be debug only

Boot Stage 2

- New boot stage 2 for `AT25SF128A`

Release 1.3.0 (02/Nov/2021)

This release contains numerous bug fixes and documentation improvements. Additionally, it contains the following notable changes/improvements:

Updated TinyUSB to 0.12.0

- The `lib/tinyusb` submodule has been updated from 0.10.1 to 0.12.0. See https://github.com/hathach/tinyusb/releases/tag/0.11.0 and https://github.com/hathach/tinyusb/releases/tag/0.12.0 for release notes.
- Improvements have been made for projects that include TinyUSB and also compile with enhanced warning levels and `-Werror`. Warnings have been fixed in rp2040 specific TinyUSB code, and in TinyUSB headers, and a new cmake function `suppress_tinyusb_warnings()` has been added, that you may call from your `CMakeLists.txt` to suppress warnings in other TinyUSB C files.

New Board Support

The following boards have been added and may be specified via `PICO_BOARD`:

- `adafruit_trinkey_qt2040`

- `melopero_shake_rp2040`

- `pimoroni_interstate75`

- `pimoroni_plasma2040`

- `pybstick26_rp2040`

- `waveshare_rp2040_lcd_0.96`

- `waveshare_rp2040_plus_4mb`

- `waveshare_rp2040_plus_16mb`

- `waveshare_rp2040_zero`

Updated SVD, `hardware_regs`, `hardware_structs`

The RP2040 SVD has been updated, fixing some register access types and adding new documentation.

The `hardware_regs` headers have been updated accordingly.

The `hardware_structs` headers which were previously hand coded, are now generated from the SVD, and retain select documentation from the SVD, including register descriptions and register bit-field tables.

e.g. what was once

```
typedef struct {
    io_rw_32 ctrl;
    io_ro_32 fstat;
    ...
```

becomes:

```
// Reference to datasheet: https://datasheets.raspberrypi.com/rp2040/rp2040-datasheet.pdf#tab-registerlist_pio
//
// The _REG_ macro is intended to help make the register navigable in your IDE (for example, using the "Go to Definition" feature)
// _REG_(x) will link to the corresponding register in hardware/regs/pio.h.
//
// Bit-field descriptions are of the form:
// BITMASK [BITRANGE]: FIELDNAME (RESETVALUE): DESCRIPTION

typedef struct {
    _REG_(PIO_CTRL_OFFSET) // PIO_CTRL
    // PIO control register
    // 0x00000f00 [11:8]  : CLKDIV_RESTART (0): Restart a state machine's clock divider from an initial phase of 0
    // 0x000000f0 [7:4]   : SM_RESTART (0): Write 1 to instantly clear internal SM state which may be otherwise difficult...
    // 0x0000000f [3:0]   : SM_ENABLE (0): Enable/disable each of the four state machines by writing 1/0 to each of these four bits
    io_rw_32 ctrl;

    _REG_(PIO_FSTAT_OFFSET) // PIO_FSTAT
    // FIFO status register
    // 0x0f000000 [27:24] : TXEMPTY (0xf): State machine TX FIFO is empty
    // 0x000f0000 [19:16] : TXFULL (0): State machine TX FIFO is full
    // 0x00000f00 [11:8]  : RXEMPTY (0xf): State machine RX FIFO is empty
    // 0x0000000f [3:0]   : RXFULL (0): State machine RX FIFO is full
```

```
        io_ro_32 fstat;
        ...
```

Behavioural Changes

There were some behavioural changes in this release:

pico_sync

SDK 1.2.0 previously added recursive mutex support using the existing (previously non-recursive) `mutex_` functions. This caused a performance regression, and the only clean way to fix the problem was to return the `mutex_` functions to their pre-SDK 1.2.0 behaviour, and split the recursive mutex functionality out into separate `recursive_mutex_` functions with a separate `recursive_mutex_` type.

Code using the SDK 1.2.0 recursive mutex functionality will need to be changed to use the new type and functions, however as a convenience, the pre-processor define `PICO_MUTEX_ENABLE_SDK120_COMPATIBILITY` may be set to 1 to retain the SDK 1.2.0 behaviour at the cost of an additional performance penalty. The ability to use this pre-processor define will be removed in a subsequent SDK version.

pico_platform

- `pico.h` and its dependencies have been slightly refactored so it can be included by assembler code as well as C/C code. This ensures that assembler code and C/C code follow the same board configuration/override order and see the same configuration defines. This should not break any existing code, but is notable enough to mention.
- `pico/platform.h` is now fully documented.

pico_standard_link

`-Wl,max-page-size=4096` is now passed to the linker, which is beneficial to certain users and should have no discernible impact on the rest.

Other Notable Improvements

hardware_base

- Added `xip_noalloc_alias(addr)`, `xip_nocache_alias(addr)`, `xip_nocache_noalloc_alias(addr)` macros for converting a flash address between XIP aliases (similar to the `hw_xxx_alias(addr)` macros).

hardware_dma

- Added `dma_timer_claim()`, `dma_timer_unclaim()`, `dma_claim_unused_timer()` and `dma_timer_is_claimed()` to manage ownership of DMA timers.
- Added `dma_timer_set_fraction()` and `dma_get_timer_dreq()` to facilitate pacing DMA transfers using DMA timers.

hardware_i2c

- Added `i2c_get_dreq()` function to facilitate configuring DMA transfers to/from an I2C instance.

hardware_irq

- Added `irq_get_priority()`.
- Fixed implementation when `PICO_DISABLE_SHARED_IRQ_HANDLERS=1` is specified, and allowed `irq_add_shared_handler` to be used in this case (as long as there is only one handler - i.e. it behaves exactly like `irq_set_exclusive_handler`).
- Sped up IRQ priority initialization which was slowing down per core initialization.

hardware_pio

- `pio_encode_` functions in `hardware/pico_instructions.h` are now documented.

hardware_pwm

- Added `pwm_get_dreq()` function to facilitate configuring DMA transfers to a PWM slice.

hardware_spi

- Added `spi_get_dreq()` function to facilitate configuring DMA transfers to/from an SPI instance.

hardware_uart

- Added `uart_get_dreq()` function to facilitate configuring DMA transfers to/from a UART instance.

hardware_watchdog

- Added `watchdog_enable_caused_reboot()` to distinguish a watchdog reboot caused by a watchdog timeout after calling `watchdog_enable()` from other watchdog reboots (e.g. that are performed when a UF2 is dragged onto a device in BOOTSEL mode).

pico_bootrom

- Added new constants and function signature typedefs to `pico/bootrom.h` to facilitate calling bootrom functions directly.

pico_multicore

- Improved documentation in `pico/multicore.h`; particularly, `multicore_lockout_` functions are newly documented.

pico_platform

- `PICO_RP2040` is now defined to 1 in `PICO_PLATFORM=rp2040` (i.e. normal) builds.

pico_stdio

- Added `puts_raw()` and `putchar_raw()` to skip CR/LF translation if enabled.
- Added `stdio_usb_connected()` to detect CDC connection when using `stdio_usb`.
- Added `PICO_STDIO_USB_CONNECT_WAIT_TIMEOUT_MS` define that can be set to wait for a CDC connection to be established during initialization of `stdio_usb`. Note: value -1 means indefinite. This can be used to prevent initial program output being lost, at the cost of requiring an active CDC connection.

- Fixed `semihosting_putc` which was completely broken.

pico_usb_reset_interface

- This new library contains `pico/usb_reset_interface.h` split out from `stdio_usb` to facilitate inclusion in external projects.

CMake build

- `OUTPUT_NAME` target property is now respected when generating supplemental files (`.BIN`, `.HEX`, `.MAP`, `.UF2`)

pioasm

- Operator precedence of `*`, `/`, `-`, `+` have been fixed
- Incorrect MicroPython output has been fixed.

elf2uf2

- A bug causing an error with binaries produced by certain other languages has been fixed.

Release 1.3.1 (18/May/2022)

This release contains numerous bug fixes and documentation improvements which are not all listed here; you can see the full list of individual commits here.

New Board Support

The following boards have been added and may be specified via `PICO_BOARD`:

- `adafruit_kb2040`
- `adafruit_macropad_rp2040`
- `eetree_gamekit_rp2040`
- `garatronic_pybstick26_rp2040` (renamed from `pybstick26_rp2040`)
- `pimoroni_badger2040`
- `pimoroni_motor2040`
- `pimoroni_servo2040`
- `pimoroni_tiny2040_2mb`
- `seeed_xiao_rp2040`
- `solderparty_rp2040_stamp_carrier`
- `solderparty_rp2040_stamp`
- `wiznet_w5100s_evb_pico`

Notable Library Changes/Improvements

hardware_dma

- New documentation has been added to the `dma_channel_abort()` function describing errata RP2040-E13, and how to work around it.

hardware_irq

- Fixed a bug related to removing and then re-adding shared IRQ handlers. It is now possible to add/remove handlers as documented.
- Added new documentation clarifying the fact the shared IRQ handler ordering "priorities" have values that increase with higher priority vs. Cortex M0+ IRQ priorites which have values that decrease with priority!

hardware_pwm

- Added a `pwm_config_set_clkdiv_int_frac()` method to complement `pwm_config_set_clkdiv_int()` and `pwm_config_set_clkdiv()`.

hardware_pio

- Fixed the `pio_set_irqn_source_mask_enabled()` method which previously affected the wrong IRQ.

hardware_rtc

- Added clarification to `rtc_set_datetime()` documentation that the new value may not be visible to a `rtc_get_datetime()` very soon after, due to crossing of clock domains.

pico_platform

- Added a `busy_wait_at_least_cycles()` method as a convenience method for a short tight-loop counter-based delay.

pico_stdio

- Fixed a bug related to removing stdio "drivers". `stdio_set_driver_enabled()` can now be used freely to dynamically enable and disable drivers during runtime.

pico_time

- Added an `is_at_the_end_of_time()` method to check if a given time matches the SDK's maximum time value.

Runtime

A bug in `__ctzdi2()` aka `__builtin_ctz(uint64_t)` was fixed.

Build

- Compilation with GCC 11 is now supported.

- `PIOASM_EXTRA_SOURCE_FILES` is now actually respected.

pioasm

- Input files with Windows (CRLF) line endings are now accepted.
- A bug in the python output was fixed.

elf2uf2

- Extra padding was added to the UF2 output of misaligned or non-contiguous binaries to work around errata RP2040-E14.

> **NOTE**
>
> The 1.3.0 release of the SDK incorrectly squashed the history of the changes. A new merge commit has been added to restore the full history, and the 1.3.0 tag has been updated

Release 1.4.0 (30/Jun/2022)

This release adds wireless support for the Raspberry Pi Pico W, adds support for other new boards, and contains various bug fixes, documentation improvements, and minor improvements/added functionality. You can see the full list of individual commits here.

New Board Support

The following boards have been added and may be specified via `PICO_BOARD`:

- `pico_w`
- `datanoisetv_rp2040_dsp`
- `solderparty_rp2040_stamp_round_carrier`

Wireless Support

- Support for the Raspberry Pi Pico W is now included with the SDK (`PICO_BOARD=pico_w`). The Pico W uses a driver for the wireless chip called `cyw43_driver` which is included as a submodule of the SDK. You need to initialize this submodule for Pico W wireless support to be available. Note that the LED on the Pico W board is only accessible via the wireless chip, and can be accessed via `cyw43_arch_gpio_put()` and `cyw43_arch_gpio_get()` (part of the `pico_cyw43_arch` library described below). As a result of the LED being on the wireless chip, there is no `PICO_DEFAULT_LED_PIN` setting and the default LED based examples in pico-examples do not work with the Pico W.
- IP support is provided by lwIP which is also included as a submodule which you should initialize if you want to use it.

    The following libraries exposing lwIP functionality are provided by the SDK:

    - `pico_lwip_core` (included in `pico_lwip`)
    - `pico_lwip_core4` (included in `pico_lwip`)
    - `pico_lwip_core6` (included in `pico_lwip`)
    - `pico_lwip_netif` (included in `pico_lwip`)

- `pico_lwip_sixlowpan` (included in `pico_lwip`)
- `pico_lwip_ppp` (included in `pico_lwip`)
- `pico_lwip_api` (this is a blocking API that may be used with FreeRTOS and is not included in `pico_lwip`)

As referenced above, the SDK provides a `pico_lwip` which aggregates all of the commonly needed lwIP functionality. You are of course free to use the substituent libraries explicitly instead.

The following libraries are provided that contain the equivalent lwIP application support:

- `pico_lwip_snmp`
- `pico_lwip_http`
- `pico_lwip_makefsdata`
- `pico_lwip_iperf`
- `pico_lwip_smtp`
- `pico_lwip_sntp`
- `pico_lwip_mdns`
- `pico_lwip_netbios`
- `pico_lwip_tftp`
- `pico_lwip_mbedtls`

- Integration of the IP stack and the `cyw43_driver` network driver into the user's code is handled by `pico_cyw43_arch`. Both the IP stack and the driver need to do work in response to network traffic, and `pico_cyw43_arch` provides a variety of strategies for servicing that work. Four architecture variants are currently provided as libraries:

    - `pico_cyw43_arch_lwip_poll` - For using the RAW lwIP API (`NO_SYS=1` mode) with polling. With this architecture the user code must periodically poll via `cyw43_arch_poll()` to perform background work. This architecture matches the common use of lwIP on microcontrollers, and provides no multicore safety
    - `pico_cyw43_arch_lwip_threadsafe_background` - For using the RAW lwIP API (`NO_SYS=1` mode) with multicore safety, and automatic servicing of the `cyw43_driver` and lwIP in the background. User polling is not required with this architecture, but care should be taken as lwIP callbacks happen in an IRQ context.
    - `pico_cyw43_arch_lwip_sys_freertos` - For using the full lwIP API including blocking sockets in OS mode (`NO_SYS=0`), along with multicore/task safety, and automatic servicing of the `cyw43_driver` and the lwIP stack in a separate task. This powerful architecture works with both SMP and non-SMP variants of the RP2040 port of FreeRTOS-Kernel. Note you must set `FREERTOS_KERNEL_PATH` in your build to use this variant.
    - `pico_cyw43_arch_none` - If you do not need the TCP/IP stack but wish to use the on-board LED or other wireless chip connected GPIOs.

See the library documentation or the `pico/cyw43_arch.h` header for more details.

Notable Library Changes/Improvements

hardware_dma

- Added `dma_unclaim_mask()` function for un-claiming multiple DMA channels at once.
- Added `channel_config_set_high_priority()` function to set the channel priority via a channel config object.

hardware_gpio

- Improved the documentation for the pre-existing gpio IRQ functions which use the "one callback per core" callback mechanism, and added a `gpio_set_irq_callback()` function to explicitly set the callback independently of enabling per pin GPIO IRQs.

- Reduced the latency of calling the existing "one callback per core" GPIO IRQ callback.

- Added new support for the user to add their own shared GPIO IRQ handler independent of the pre-existing "one callback per core" callback mechanism, allowing for independent usage of GPIO IRQs without having to share one handler.

  See the documentation in `hardware/irq.h` for full details of the functions added:

  - `gpio_add_raw_irq_handler()`
  - `gpio_add_raw_irq_handler_masked()`
  - `gpio_add_raw_irq_handler_with_order_priority()`
  - `gpio_add_raw_irq_handler_with_order_priority_masked()`
  - `gpio_remove_raw_irq_handler()`
  - `gpio_remove_raw_irq_handler_masked()`

- Added a `gpio_get_irq_event_mask()` utility function for use by the new "raw" IRQ handlers.

hardware_irq

- Added `user_irq_claim()`, `user_irq_unclaim()`, `user_irq_claim_unused()` and `user_irq_is_claimed()` functions for claiming ownership of the **user** IRQs (the ones numbered 26-31 and not connected to any hardware). Uses of the **user** IRQs have been updated to use these functions. For `stdio_usb`, the `PICO_STDIO_USB_LOW_PRIORITY_IRQ` define is still respected if specified, but otherwise an unclaimed one is chosen.

- Added an `irq_is_shared_handler()` function to determine if a particular IRQ uses a shared handler.

pico_sync

- Added a `sem_try_acquire()` function, for non blocking acquisition of a semaphore.

pico_stdio

- `stderr` is now supported and goes to the same destination as `stdout`.
- Zero timeouts for `getchar_timeout_us()` are now correctly honored (previously they were a 1µs minimum).

stdio_usb

- The use of a 1ms timer to handle background TinyUSB work has been replaced with use of a more interrupt driven approach using a **user** IRQ for better performance. Note this new feature is disabled if shared IRQ handlers are disabled via `PICO_DISABLE_SHARED_IRQ_HANDLERS=1`

Miscellaneous

- `get_core_num()` has been moved to `pico/platform.h` from `hardware/sync.h`.
- The C library function `realloc()` is now multicore safe too.
- The minimum PLL frequency has been increased from 400Mhz to 750Mhz to improve stability across operating conditions. This should not affect the majority of users in any way, but may impact those trying to set particularly low clock frequencies. If you do wish to return to the previous minimum, you can set `PICO_PLL_VCO_MIN_FREQ_MHZ` back to `400`. There is also a new `PICO_PLL_VCO_MAX_FREQ_MHZ` which defaults to `1600`.

Build

- Compilation with GCC 12 is now supported.

Appendix G: Documentation release history

Table 27. Documentation release history

Release	Date	Description
1.0	21 Jan 2021	- Initial release
1.1	26 Jan 2021	- Minor corrections - Extra information about using DMA with ADC - Clarified M0+ and SIO CPUID registers - Added more discussion of Timers - Update Windows and macOS build instructions - Renamed books and optimised size of output PDFs
1.2	01 Feb 2021	- Minor corrections - Small improvements to PIO documentation - Added missing TIMER2 and TIMER3 registers to DMA - Explained how to get MicroPython REPL on UART - To accompany the V1.0.1 release of the C SDK
1.3	23 Feb 2021	- Minor corrections - Changed font - Additional documentation on sink/source limits for RP2040 - Major improvements to SWD documentation - Updated MicroPython build instructions - MicroPython UART example code - Updated Thonny instructions - Updated Project Generator instructions - Added a FAQ document - Added errata E7, E8 and E9
1.3.1	05 Mar 2021	- Minor corrections - To accompany the V1.1.0 release of the C SDK - Improved MicroPython UART example - Improved Pinout diagram
1.4	07 Apr 2021	- Minor corrections - Added errata E10 - Note about how to update the C SDK from Github - To accompany the V1.1.2 release of the C SDK

Release	Date	Description
1.4.1	13 Apr 2021	- Minor corrections - Clarified that all source code in the documentation is under the 3-Clause BSD license.
1.5	07 Jun 2021	- Minor updates and corrections - Updated FAQ - Added SDK release history - To accompany the V1.2.0 release of the C SDK
1.6	23 Jun 2021	- Minor updates and corrections - ADC information updated - Added errata E11
1.6.1	30 Sep 2021	- Minor updates and corrections - Information about B2 release - Updated errata for B2 release
1.7	03 Nov 2021	- Minor updates and corrections - Fixed some register access types and descriptions - Added core 1 launch sequence info - Described SDK "panic" handling - Updated picotool documentation - Additional examples added to **Appendix A: App Notes** appendix in the Raspberry Pi Pico C/C++ SDK book - To accompany the V1.3.0 release of the C SDK
1.7.1	04 Nov 2021	- Minor updates and corrections - Better documentation of USB double buffering - Picoprobe branch changes - Updated links to documentation
1.8	17 Jun 2022	- Minor updates and corrections - Updated setup instructions for Windows in Getting started with Raspberry Pi Pico - Additional explanation of SDK configuration - RP2040 now qualified to -40°C, minimum operating temperature changed from -20°C to -40°C - Increased PLL min VCO from 400MHz to 750MHz for improved stability across operating conditions - Added reflow-soldering temperature profile - Added errata E12, E13 and E14 - To accompany the V1.3.1 release of the C SDK

Release	Date	Description
1.9	30 Jun 2022	- Minor updates and corrections - Update to VGA board hardware description for launch of Raspberry Pi Pico W - To accompany the V1.4.0 release of the C SDK
	Pico and Pico W databooks combined into a unified release history	
2.0	01 Dec 2022	- Minor updates and corrections - Added RP2040 availability information - Added RP2040 storage conditions and thermal characteristics - Replace SDK library documentation with links to the online version - Updated Picoprobe build and usage instructions

The latest release can be found at https://datasheets.raspberrypi.com/pico/raspberry-pi-pico-c-sdk.pdf.

Raspberry Pi is a trademark of Raspberry Pi Ltd

Raspberry Pi Ltd

This page was intentionally left blank.

This page was intentionally left blank.

Printed in Great Britain
by Amazon